Who Cares?

Or why war, poverty, environmental destruction and debt remain so popular

Richard Walker

Paper Coin Publishing

First published in the United Kingdom in 2011 by Paper Coin Publishing

ISBN 978-0-9568208-0-8

Produced by
The Choir Press, Gloucester
www.thechoirpress.co.uk

Who Cares?

Contents

1

Easter Lessons

A thousand years ago two ocean-going canoes set out from one of the easterly Polynesian islands around Mangareva. For around two weeks they made their way eastwards across the Pacific until they were able to follow the hunting seabirds to their huge nesting colonies on the island that we now call Rapa Nui or Easter Island. After a thousand years of colonising the islands of the Pacific man had found one of the most isolated pieces of habitable land on the planet.

The island was a dense forest of enormous trees including a species of palm tree that dwarfed any other on the planet. The trees that the Polynesians used to build their great canoes were so large and in such abundance that they called the highest mountain on Easter 'Terevaka', 'the place to get canoes'. With seemingly limitless supplies of timber to build their canoes they were able to harvest the ocean for hundreds of miles around. The forest also provided a super abundance of wild food. This was a paradise that came with a remarkable benefit: a total absence of competition from any other group of humans.

For more than two hundred years they lived well. Land was cleared for farming and as the population flourished the island was divided into separate territories. As in other Polynesian communities each clan was divided into the elite who were made up of the chiefs supported by their priests, and the rest who divided up the toil between them.

Eventually a fierce competition developed. The clans began to compete in producing the biggest stone statues and then hauling them from the island's quarry to a prominent position looking out over the ocean.

This was no mean feat. These statues were massive and dragging them through dense forest wasn't easy. Trees and shrubs had to be cleared and then a timber roadway created. The roadway to transport the statues was similar to the one used to move their newly built canoes from Terevaka to the ocean. The roadway was constructed using those same trees from which they fashioned their giant canoes.

They went at their work with enthusiasm so that by the time the first western sailors discovered the island on Easter Day 1722 the Polynesians had cleared their island of everything that could be called a tree.

This had a profound effect on the islanders who not only lost out on the abundant food that the island had once provided but they also found themselves without the trees with which to replace their old canoes. As a result deep sea fishing, or even escape to another island, was impossible. Conditions on the island in the eighteenth century were so poor that for a long time western visitors were at a loss to explain how the statues had got there. It seemed impossible that these enormous figures could have been quarried, sculptured, transported and erected by any ancestors of the poor Polynesians who were now scratching out a living on a degraded island.

There is a lesson for our global society here. As Jared Diamond points out in his excellent book *Collapse: How Societies Choose to Fail or Succeed*:[1]

> Easter's isolation makes it the clearest example of a society that destroyed itself by overexploiting its own resources. ... The parallels between Easter Island and the whole modern world are chillingly obvious.

At the end of the book he offers some reasons to suggest that we can avoid their fate. The problems we face are not insoluble.

> Because we are the cause of our environmental problems, we are the ones in control of them, and we can choose or not choose to stop causing them and start solving them. ... We don't need new technologies . . . we just need the political will to apply solutions already available.

As he goes on to say though, 'that's a big just'.

Most people today, especially those in the environmental movement, would say that it is a very big 'just'.

I should perhaps acknowledge here that there are people who say that the environmentalists are just trying to frighten us. They point out that the planet has been through a lot worse than we can throw at it: volcanoes, earthquakes, comets and temperature extremes beyond anything our environmentalists are warning us about.

That is all true. The issue though is not whether the planet will survive; it is whether we will. As the American comedian George Carlin put it, 'The planet isn't going anywhere. We are.'

He also said, 'Save the Planet! We haven't even learned to take care of each other!' That is a very interesting point. It would seem obvious that if the Easter Islanders had really been taking care of each other then the island might not have suffered quite so much loss of diversity. After all, taking care of each other must include taking care of each other's living space.

It was not inevitable. Because, even though there were factors that the Easter Islanders had no control over, there were options available to them that would have allowed them to avoid their fate. They certainly could not have been blind to what was happening. The deforestation didn't happen overnight, or even over just one or two generations.

Even in the early stages it is extremely unlikely that absolutely nobody anticipated the enormous problem that they were creating. Perhaps some older woman, observing the glamorous and probably male world of the quarry and the doubtlessly well rewarded construction teams competing to complete the latest timber roadway to the sea, saw a catastrophe in the making. Working at the less prestigious end of the economy, tending vegetables or gathering wild food, her perception might have been different from the elite power-brokers. She might have suggested that their activities, while possibly appealing to some god somewhere, would effectively dig the islanders into a hole from which it would be impossible to escape.

I would guess that the elite's response would have varied between mockery and angry indignation. The elite were of course the gods' representatives on earth. Hadn't they proved that? Who but their friendly gods would have led them to this paradise?

Once half the island had been stripped of trees the problem presumably became visible to a wider section of the population. Even the gangs working on replacing, repairing and extending the canoe ladders might have begun to wake up to the looming crisis.

Perhaps a Thomas L. Friedman among them might

have come up with a Polynesian equivalent of *Hot, Flat, and Crowded.*[2] He might have, like our present-day Thomas, have come out with some neat technological solutions to the problem. He might have suggested a method of construction that uses 10 or 20 per cent less timber. Voluntary quotas on trees felled – although not so much that the Easter economy would go into a tailspin because the gods felt insulted.

Some Vince Cable might have pointed out that a loss of trees was inevitable but that if proper regulations had been put in place then the Easter Islanders could have probably weathered the storm. They could have regulated the amount of timber used in cooking fires or got tougher on the loggers about waste and inefficiency.

Other less exalted individuals in the community might have said, 'Okay, big deal, there's a problem, we are losing trees but I've got a job at stake so, sorry, I'm not going to start worrying about trees.'

Another might have pointed out that some Tahitians had recently been blown off course and that these 'immigrants' had been allowed to stay taking food, timber and jobs. Somebody might have come up with the idea of a 'big society' where people volunteered to go out planting replacement trees which in 200 years would ... So symptoms would have been identified and remedies offered.

It is just possible that somebody went beyond the symptoms and managed to identify the cause of the problem. They might have said, 'Why are we destroying the island just to get these enormous statues in place? The chiefs and their priests say it's what the gods expect. Who are these gods? Why do they want us to work so hard to destroy the home that they apparently gave us? Maybe the priests have got it wrong. Or maybe they even made up these stories about the gods and what they want because it keeps the chiefs and priests in a position of power.'

A look of horror might pass across the faces of her listeners as they admonish her for her ridiculous notions. The woman might persist, 'Surely you can see that the real cause of our problems is the hugely wasteful and, I have to say, totally useless economic endeavour of constructing these interesting but non-essential statues?' (Well that's how she would put it if they interviewed her on Easter Island's *Woman's Hour*.)

Some of the women might say to her, 'Stop it! We have a good life here, at least we are free to say what we like and enjoy ourselves. Our chiefs are good and caring. Stop trying to make trouble.'

Somebody might even accuse her of inventing some ridiculous conspiracy theory. 'You really think that they would make up a story like that? No! The simple explanation is the right one. The gods want us to put these big idols on the cliff top.'

So if you suddenly ended up on Easter Island 400 years ago what would you do? You might tell them that they are on track to destroy their island within two or three generations. You might tell them to stop wasting trees for a belief system that has become destructive. You could tell them that they just need to find the political will to worship their gods in a less destructive manner; but would they listen to you?

The important question is why they stubbornly held onto belief systems that were long past their sell-by date. This is important because the Polynesians on Easter Island were anything but unique. Throughout human history societies have gone to enormous lengths just to be able to continue with systems that have become toxic. The Easter Islanders were no different from every other culture including our own. And now that the world is a global village we, like the people of Easter Island, have nowhere to escape to if our toxic belief systems lead us to degrade the planet in the same way that the Easter Islanders degraded the island paradise that they discovered.

2

Beyond Belief

Belief is powerful. And you do not need to be a religious person to know that. Every track athlete, tennis player and footballer knows the part that belief plays in winning. But sometimes we have to check our beliefs to make sure that we are staying in touch with reality.

There appears to be considerable evidence around us to suggest that we need to change some of our fundamental belief systems but as psychologists know we are prepared to die first. They explain our tenacious desire to hold onto beliefs, even when all the evidence contradicts what we believe, as a survival tool.

For most of our hundreds of thousands of years of evolution we were hunters. Beliefs as to the most effective way to perform tasks were passed on. Suddenly faced with an angry mammoth and surrounded by your lightly armed companions was not the time to start considering a new paradigm for dealing with the angry mammoth no matter how good the idea might appear.

As children we learn to obey without question – a useful survival tool when mum tells us to stay away from that sabre toothed tiger.

This behaviour is useful but not when circumstances have changed so much that our old belief systems are positively dangerous. There are numerous cases of societies dying because of an inability to change the way they did things as circumstances altered. This was a catastrophe for those in the society but not for mankind as a whole. Indeed some catastrophes for one society may have turned into an opportunity for another. The problem today is that there is no other society.

Another issue that makes it difficult to change is that once a system starts to get going the most able and ambitious in the society naturally fill the key places. They are the ones we look to for leadership and generally they are equipped for the task. Unfortunately though they are also the ones who receive the biggest rewards from the current system and they will therefore tend to use all their abilities and skills to keep the existing system in place. So the obvious problem is that when our belief system needs changing the very personalities who are best able to help us move on are in fact using every weapon in their armoury to defend that belief system.

Also the system itself will tend to keep the best brains very busy so that they will lack the time to contemplate a new way of operating.

And even if our best and brightest can overcome self interest and a demanding work schedule there is the problem of finding the immense courage that they will require if they are to be among the first to bring in a new order.

Distraction is also useful in defending the existing system. You only have to look at party politics over the last fifty years to see what a multitude of distractions have been created rather than deal with the problem at its root. And of course for all of us there are countless distractions. Television, sport, alcohol, film and theatre; indeed these distractions are essential sometimes to help us deal with the stress caused by the society that has grown up around us.

In short there are countless reasons why we, like past societies, have been unable to 'just' find that political will.

Of course we need not get dispirited. Just because we have a tendency to hang on to obsolete belief systems it does not mean we always will. Indeed just knowing about the tendency can have a positive effect. And there have been instances where apparently impossible new beliefs have replaced cherished old beliefs.

Common sense told countless generations back through eons of time that the sun went around the earth. It was a clearly observable fact reinforced by the wisdom of kings and priests. It was a fact backed up by science as navigators used complex mathematics, combined with an assumption that the heavenly bodies went around the earth, to find their way across great tracts of ocean. By what possible means then could such a 'sensible' view of the order of things be overturned? Well of course it did take a few hundred years but it did happen. Few will have any problem with the belief that the earth travels around the sun.

So deeply rooted belief systems can change; however there do have to be certain conditions in place. First there need to be compelling reasons to believe that change is necessary. Then there needs to be enough time for the change to take place. And crucially there needs to be the right combination of people to bring about the change. On Easter Island it is possible to believe that eventually the right combination of courageous individuals among the chiefs or priests might have come along. Sadly time ran out. The first condition was certainly in place. There were most definitely compelling reasons for change.

Do *we* have compelling reasons for change? I imagine

some people will think so. The question is, how compelling? Compelling enough for the mass of people to realize that they can't just leave it to the chiefs, then shrug and mutter, 'After all what can I do?' Compelling enough for the chiefs to find the courage to turn the spotlight on the real cause of our problems rather than fiddling with the symptoms?

3

Compelling Reasons

I am going to list what I believe are the compelling reasons for our society to change its deeply held belief systems.

1) The economy is at war with our environment

We have an economic system that all key decision makers believe is more important than anything else. Our 'economy' definitely takes priority in any decision making. As the Clinton election slogan swaggeringly put it, 'It's the economy, stupid.'

Just as the Easter Islanders felt they must sacrifice everything to get more and more statues in place so we must sacrifice everything, including the environment, to keep our economy satisfied.

2) Poverty is massive and growing

Poverty will always be with us. Why? We raise millions in charity appeals, so somebody cares. The question is do we care enough?

The Nobel Prize winner and ex-Chief Economist of the World Bank, Joseph Stiglitz, has calculated that 3 trillion dollars and 2 million men have been used to fight a single war against Iraq. With those enormous resources focused on eliminating poverty it would definitely have been history by now.

If you are thinking, 'Ridiculous! Life is not like that,' then you must come up with a very clear-headed explanation as to why it isn't like that. We obviously have the ability to eliminate poverty so why haven't we?

3) War

While technology has been delivering its bounty, wars have been fought in greater numbers and with greater devastation than at any time in history. Surely any sane person must believe that by now we should have found another way to resolve issues than by devastating whole countries and slaughtering civilian populations.

4) Spending Cuts

Okay, after environmental devastation, starvation in the midst of massive abundance, and of course war, spending cuts may seem positively trivial but I want you to ask yourself a question and then allow time to really consider the answer. If things get really tough and you realize that some kind of savings just have to be made what would your first choices be?

Would you pounce on your child's education? Or your family's healthcare? Or would you say, 'We'll wait for that new car. That new mobile phone, that holiday abroad.' Think about that. Now remind yourself what you are being told is the sensible solution to our difficulties.

5) Boom and Bust

There is no doubt that what is politely referred to as 'the business cycle' can cause terrible suffering. People recently have lost their homes. Businesses, that have taken a lifetime to build, have been destroyed overnight. Even pensions, earned through years of toil, have magically evaporated. Why has that happened? And why will it keep happening?

In most areas of endeavour we try to work logically. We generally believe that things happen for a reason. If we don't like a certain outcome we try to find the cause and then do something about it. In an area like the economy that has such a huge impact on every single human being we should be doing better. This is the economy, not nuclear physics, brain surgery or rocket science. The economy is simply about connecting people with the goods and the services that we produce. It might be bigger but it has got to be infinitely less complicated than designing an iPhone.

Five issues that I think are very serious. At least one of them has to be important to you.

Maybe you are passionate about the rainforests, the coral reefs, the blue whale or even just the world that your children and grandchildren will inherit.

Maybe you are desperately trying to feed your kids and pay the rent and still find money for your railcard to get you to your place of work.

Or possibly you don't want to sit through another famine relief programme on BBC Two while your kids ask difficult questions.

Maybe you don't want to hear any more lies from a politician desperate to invade some impoverished but resource-rich country.

Maybe you don't want to leave university saddled with a huge debt. Or maybe you want your elderly parent to have a dignified end to their life.

Maybe you just don't want your hairdressing business, your chip shop or your decorating firm to go belly up in the next banking crisis.

You have to be in there somewhere.

4
Your Choice

Think for a moment of your favourite explanation for what is wrong in all five areas. Say it to yourself. Too many greedy people, too many immigrants, too many scroungers, not enough government support for the poor or for our domestic industry, not enough support for private enterprise, too much power handed to the European Union, too much waste. Not enough genuine socialism. Not enough genuine capitalism. Not enough patriotism. Not enough democracy. If I've missed out your explanation I apologise and do please email me about it at thepapercoin@gmail.com.

You probably didn't find that too difficult but now I want you to put that favourite explanation behind you. Not for ever, just for the next half-hour or so. Pretend that you think all of those explanations, including your favourite one, have been argued over for generations and we haven't really got anywhere.

Pretend that you believe that 300 years of massive technological advancement should have created a healthier planet and a more compassionate society. Not Utopia, we are talking about human beings here. Utopia will never exist. No, I just want you to imagine a world where people do a lot more than have the occasional whip-round to stop those pictures of swollen bellies and emaciated mothers that flick across our screens from time to time.

Okay so you travel back in time and end up on Easter Island a few hundred years ago and somebody arrives from another galaxy and suggests that you persuade your friends, your chiefs and your priests that you need a different relationship with your gods – would you listen to them? Would you be able to say that you will give up pleasing the gods by presenting them with giant idols?

Not too difficult I shouldn't think. You might even think it was just plain common sense to stop destroying the forests for the sake of some pagan myth. If you are honest you might find that you have a pretty low opinion of the noble savages on that island. You might feel that only their incredibly primitive thinking could make them chop down trees with such reckless abandon, especially when it was for no truly useful purpose. Indeed you might think that you would never support such foolish behaviour. You might; but then you weren't brought up being conditioned by Polynesian belief systems.

Think about it. You would be asking these Polynesians to throw away the life they have known since birth. They would have to learn a whole new set of beliefs. So what would you do? Shake your head and say, 'No, this will be far too painful for these people. Change might kill them. Let them carry on the way they are until all the trees have gone.'?

If you were there on the island with your family what would you do? Would you say, 'This is going to be tough and I am not going to be popular but I have got to fight tooth and nail to get them to change their beliefs. If they keep on the way they are they will destroy so much of what they have. My children and grandchildren are going to be badly affected.'?

It's your choice; what would you do?

5

Money Changes Everything

How would you react if somebody showed you a single cause for all five areas of concern that I have listed? Would you say, 'No that's ridiculous; a single cause? No it's much more complicated than that.' Or, 'If it was that simple our chiefs would have realized it. They are not stupid you know.'

If I asked you what was the single most important issue that has affected and affects your life you might say your health, friendship, love or even your job. If I said what tangible thing in the world has had the most significant impact on all those areas you might mention money. If you look at all five issues mentioned I think you will agree that money has to play a part in all five areas.

You don't have to have read Niall Ferguson's *The Ascent of Money*[1] to know that money is one of the most powerful tools that man has created. Today money is so much a part of our everyday existence that we pay it little conscious thought. We certainly put a lot of thought into making sure we get hold of it and we are certainly aware of how miserable life can be without it. We certainly think about the effect money has on us at a personal level, but what about the wider impact of money?

Money has had a very powerful effect on our history, far more than most of us realize. In *The Ascent of Money* Niall Ferguson says that in the eighth and ninth centuries 'there was a chronic shortage of silver in Western Europe. Demand for money was greater in the much more developed commercial centres of the Islamic Empire' and so that is where the silver money drained to. This seriously hampered Western Europe's trade.

He says that the Europeans tried to solve this problem in two ways. One way was by 'exchanging slaves and timber for silver in Baghdad or for African gold in Cordoba and Cairo'. The other way was to 'plunder precious metal by making war on the Muslim world'. He says that, 'The Crusades, like the conquests that followed, were as much about overcoming Europe's monetary shortage as about converting heathens to Christianity.'

It seems reasonable to assume that the conquests might have been *all* about monetary shortage – but 'converting heathens to Christianity' had a slightly nobler ring to it.

However what Professor Ferguson's book does very clearly demonstrate is just how much our history has been shaped by money.

6

Money Matters

There is of course no question about what a useful tool money is. Without money we would be forced to carry whatever we have made and try to swap it for what we want.

Let's take a simplified example. An island contains a hundred people all working industriously. A carpenter finishes a chair and carries it to the baker's and swaps it for 50 loaves of bread. The problem is obvious.

Now somebody suggests that everybody is given tokens. We'll call them pounds. Each of the hundred people on the island is given one thousand pounds. Now the carpenter goes down to the baker's and hands two of his tokens or two of his pounds to the baker and takes his loaf. The baker goes to the seamstress and hands over five tokens and walks off with a shirt. Some days later the seamstress hands 50 tokens to the carpenter and walks off with a chair.

Now what is important in this example is that at the end of a year the same amount of money will still be moving around. No more money has had to be added. They can keep circulating that money for ever.

Now imagine another island with 100 people all swapping goods. Now an outsider arrives and says, I can make life a whole lot easier for you. He shows them some tokens and tells them it's called money and he explains how it works. The hundred people all see that this has advantages.

The man says, I will let you have 1000 tokens each. They say that sounds good. Obviously, he says, I will need you to sign a piece of paper saying that you have taken 1000 of my tokens that will make your life a whole lot easier. You can pay that back over the next ten years. Oh, I will have to make a tiny charge. Obviously I can't do this for nothing. So you will pay 8% on top of what you owe for the next ten years. Obviously this is going to be good for you because this is going to make it much easier to operate your business and you will have no problem paying me back because you are going to be earning money from other people. And of course if you hit a sticky patch come back to me and I'm sure we can arrange something.

With the first example the money would circulate for ever. With the second example the money would need to be topped up. Everybody would be returning their loan in instalments back to the man – let's call him the banker – with interest. The banker might point out that this is a much better way to run the 'economy' because people are going to be competing at a much higher level than when the money was simply circulating. He would be right. The more money you take from everybody else the easier it is to make your repayments.

If you are efficient and hardworking you can make your payments easily and have money left over. Now if you have money left over somebody else is not going to have enough to make his payments. Obviously; because the banker only put so much into the pool and out of that pool his payments have to come.

The successful people will be able to point out that it is usually the less industrious or foolish in society that will end up without any money. This may be true but that is not the reason that they will end up with nothing. Even if there were no work-shy or incompetent people there would still be a lack of money for some people. It is impossible for this particular system to work in any other way. No matter how viciously people fight over the money there will not be enough to allow people to trade with each other and repay the banker. It is a simple matter of mathematics.

Eventually some people are going to be going to the banker saying, 'I'm sorry but I can't buy everything I need and make my monthly payment to you because there just doesn't seem to be enough money.' Well that's easy to solve the banker says, just borrow more money. In fact everybody borrow more money and we will get this economy moving.

More money has been borrowed and for a while things go well but then again those who are not selling that must-have-tastes-wonderful-and-makes-me-forget-my-worries coconut juice will realise that their borrowed money is again racing away from them. A lot of it of course back to the banker, a nice man who keeps solving the problem by giving out more money.

Eventually of course his patience will run out and he will say, 'Look you're just living beyond your means. I think the only thing to do is for you to sell me your business which will help you pay some of what you owe me and maybe sell me your house and that will help as well. Of course times are bad so the value of your house and business has dropped. Still I'll pay you enough so you won't owe as much to me. You can now stay in the house which is now mine and pay me rent and to help you pay that rent I will let you continue what you were doing and I will pay you a wage. And I will bring in my friend who has made a killing on that magic coconut juice that you enjoy because it helps you forget all your worries and he will manage this business more efficiently and get you turning a better profit which will make me very happy.'

Okay, can you see how the first system is going to bump along for ever on the same amount of money? The second system using a banker who lends the money to everyone won't bump along for ever circulating the same money. The money has to be paid back. If everybody did make their repayments eventually the banker would end up with all the money and the people would either be back swapping their goods or they would have to borrow more money.

Now the best and the brightest, and possibly the unscrupulous, are going to start looking for the work or the racket that helps them end up with the biggest share of the money so at least they will be able to make the repayments to the banker and have enough to live on. If one of these bright and dynamic individuals finds a really lucrative area of the economy to work in he might end up getting even more of the money than that. Now obviously if he gets a much bigger share of the money other people are going to have a much smaller share. They will not be able to make their repayments and of course they will risk losing their business and their home either to the banker or the person who has found that particular niche in the market that makes that big return.

7

'You Have To Be Joking!'

Yes, you are absolutely right to be sceptical. The idea that all our money is just created by a banker as debt is bizarre. You are probably saying to yourself that this is all a nice simple explanation but our system isn't like that. You are probably saying to yourself, 'Ridiculous; nobody would be stupid enough to allow a system like that to happen. Would we really let bankers, or anyone else for that matter, get away with such an obvious confidence trick. Heck if we fell for that one we would be dumber than those Wallys who hacked down every tree on their island just to make the great Umba Wumba feel good. Obviously I can see that if somebody really did get a dumb idea like that up and running all kinds of crazy things could end up happening on planet earth. It's not true. It is not the way our banking system works.'

Let me then offer some expert witnesses.

Graham Towers, Governor of the Bank of Canada from 1935 to 1955, said:

> Banks create money. That is what they are for . . . The manufacturing process to make money consists of making an entry in a book. That is all . . . Each and every time a Bank makes a loan . . . new Bank credit is created – brand new money.[1]

Robert B. Anderson, Secretary of the Treasury under President Dwight D. Eisenhower, put it this way:

> When a bank makes a loan, it simply adds to the borrower's deposit account in the bank by the amount of the loan. The money is not taken from anyone else's account; it was not previously paid in to the bank by anyone. It's new money, created by the bank for the use of the borrower.[2]

And the eminent economist, former Paul M. Warburg Professor of Economics at Harvard University, John Kenneth Galbraith simply said:

> The study of money, above all other fields in economics, is one in which complexity is used to disguise truth or to evade truth, not to reveal it. The process by which banks create money is so simple that the mind is repelled.[3]

Okay, perhaps you say, fine but I'm not going to believe anything a banker says or some government bureaucrat and definitely not some economist. Well what about a scientist and a pretty good one?

> If our nation can issue a dollar bond, it can issue a dollar bill. The element that makes the bond good makes the bill good. ... It is absurd to say that our country can issue $30,000,000 in bonds and not $30,000,000 in currency. Both are promises to pay; but one promise fattens the usurer, and the other helps the people. If the currency issued by the Government were no good, then the bonds issued would be no good either. It is a terrible situation when the Government, to increase the national wealth, must go into debt and submit to ruinous interest charges at the hands of men who control the fictitious values of gold. . . . Look at it another way. . . who is behind the Government? The

> people. Therefore it is the people who constitute the basis of Government credit. Why then cannot the people have the benefit of their own gilt-edged credit? (Thomas Edison)[4]

The way that money works has a huge effect on the planet and yet few of us give it any thought. We might think about how to get hold of more money or how to spend less money but we don't spend much time thinking about how it is created, or who decides how much of it will be created and how they decide where the money will go.

Of course there are people who have spent a lot of time thinking about just that sort of thing. They have reaped enormous benefits from this system. For those few it is not the destruction, inefficiency or injustice created by the system that matters; it is the profits.

There are also a significant number of people who depend on the system for favours. They will also do well. Priests, politicians, journalists and academics will occasionally discover that given the correct approach the system can show them warm generosity.

The great majority of our species will not show much interest and by not showing much interest they will of course be unable to see just how damaging the system really is to them. They will quietly continue shouldering the burden for those in the know.

There is, in my opinion, nothing sinister in all of this. This is how many human societies have operated including the Easter Islanders. I am not writing this to inflame tempers or to alert people to some malevolent conspiracy. I am writing to encourage a fresh look at our situation.

We have had five thousand or so years of civilisation.

Complex societies have thrived and collapsed because, inevitably, those societies have become elite-led power structures. The elite quite rationally have always put maximum effort into maintaining their system rather than adapting to changed situations. This has often led to their demise. Do read *Collapse* as Jared Diamond shows this brilliantly.

I am writing because there is no other society here to take our place. I am writing because as Jared Diamond says 'We are the cause of our environmental problems, we are the ones in control of them, and we can choose or not choose to stop causing them and start solving them.' And that doesn't just apply to our environmental problems. So can we find the necessary political will?

The realist in me hears the voice of George Carlin berating the human race for its stupidity, greed and arrogance. The planet, he snarls in that magnificent rasp of his, will shake us off like a bad case of fleas.

A small voice though whispers somewhere inside me that change can happen. However remote that hope may be, we can choose a new direction. I do not hold any religious belief but now I would like to believe that there was something in that prophecy about the meek inheriting the earth. And before anybody sneers at the impossibility of such Utopian nonsense let me ask this: isn't that prophecy a reasonable definition of democracy?

I think it is time that the meek began to take steps to understand how our money works. I think it would be useful if they ceased to be disinterested in understanding just what a tremendous advantage a few may gain from controlling a system where money can be made out of money.

Please don't think this is about greedy bankers, as

tempting as that is. The cause is the system. The bankers in the headlines are just the frontmen and some of them become the fall guys.

Every great mind that has studied our money system from J. K. Galbraith to Niall Ferguson agrees that money is made out of nothing. Which is absolutely no problem if money is used simply as a circulating token. Unfortunately in our world money is not just a circulating token.

In our world it is a philosopher's stone. Like the philosopher's stone money can turn something that has no value into something that has great value. It is magic. Unfortunately like all magic it is a trick.

Money cannot create anything out of nothing no matter how much interest a wizard charges. Somebody pays. Usually the meek. Right now the meek payers include just about every other living creature on planet earth.

I would think this a tragedy if it weren't for the fact that there is a much bigger tragedy. It doesn't have to be like this.

In the next chapters I want to show how our money affects all the five areas of concern.

8

Economy versus Environment

In which of the two systems described in the previous chapter will competition be higher? No question about it, if you owe the banker money you are going to fight tooth and nail to make a buck. If you are among the few successful ones you are going to praise this system as an incredibly efficient motivator. If somebody comes along and says please don't chop down those trees as there is an endangered species of gibbon that only lives in that forest you will point out that you have to find money to pay the banker and money to pay your workforce. Now of course, you might be a philanthropist so you could offer to start a fund to help the environment. You could say that you're happy to put half of one per cent of your profits into it but business is business, and gibbon or no gibbon, those trees have got to go.

In the system where money is not loaned into the economy but is simply put in and left to circulate I think you will agree there is going to be less pressure on individuals to chase money at all costs. All costs in this instance will include costs to the environment.

There will be people in both systems who are greedier than other people and some who are hungrier for power. Neither system will run perfectly. But there would be a great deal more pressure on you to pump out pollution, providing it makes a profit, if you borrowed a billion to get yourself into the oil industry and you know that the banker is waiting to get his money back.

Now think of your favourite solution to the environmental problems created by our economy. Now try to imagine that solution operating in each of the two systems.

Will your cleaner technology solution get a fairer shot at solving any environmental problems in the system that Thomas Edison was describing? The system where the money is put into the economy as circulating tokens to be passed around? Or will it have a better chance of making a difference where the money is constantly being passed back to the banker?

Remember that in the real world the presidential election campaign between John McCain and Barak Obama was to be the first election campaign that would be fought over the environment.

It is hard to see how anything could be more important than the environment. And I don't care what you think about global warming. Even sceptics of global warming, like David Bellamy, aren't campaigning to rip out virgin rainforest or recommending that we concrete over the Lake District. Living space is important whoever you are.

But in the middle of the election campaign along came the crash or the storm or economic meltdown if you prefer and the two candidates flipped. The economy had to pick up at all costs. Even if that cost came from the environment. The economy is king and it has to grow. And not just any part of the economy. No it had to be the important parts of the economy. Not more teachers or doctors or forestry workers or care-workers but real workers – wealth creators. In our system the economy trumps everything. The bankers must be paid back.

All the great efforts to protect the environment are fine and noble but in the face of the terrifying juggernaut that is sometimes called our economic system, they are wasted. It is like half a dozen mice trying to hold back a charging rhinoceros.

We are not destroying our planet because businessmen are incapable of running efficient sensible businesses or because we are greedy or because science is unable to find clean ways for us to survive. We are destroying it because like the Easter Islanders, we can't break from our toxic belief system. They needed to realize that there were other ways to worship their gods and we need to realize that there are much more intelligent ways of working with money.

Understanding what a powerful effect money has on our world is vital if we really want to solve our problems. I am certainly not being at all original in saying that. More than 200 years ago the second President of the United States of America, John Adams, wrote:

> All the perplexities, confusion and distresses in America arise not from defects in the constitution or confederation, nor from want of honor or virtue, as much as from downright ignorance of the nature of coin, credit, and circulation.[1]

This intelligent man is pointing out that it isn't because of our inability to organize a sensible way to govern ourselves; or because we lack compassion and understanding for one another; or because we are lazy or selfish that things have gone wrong. No it is simply because we have no understanding of the way our money system works.

9

Rich and Poor

If you read the *Evening Standard* you will know that they ran a laudable campaign to raise a £1 million to ease the poverty in London. Read the last six words again: 'to ease the poverty in London'. Not London in 1710 or London in 1810 but today. This campaign was run to ease the poverty that exists in a city that has been one of the half-dozen richest cities in the world for more than 300 years. Go back to the list and find your favourite explanation for the five issues I mentioned and see if it explains that bizarre fact.

The *Evening Standard* of 27 July 2010 reported that 'according to the think tank London Poverty Profile, 44 per cent of children in inner London, some 260,000 in total, live in low-income families'.

The *Evening Standard* of 29 July 2010 in an article written by City Editor Chris Blackhurst celebrated the fact that 'Topshop boss gives £100,000 to our appeal to help London's poor. Fashion chain billionaire Sir Philip Green has made one of the biggest donations to the Standard's Dispossessed Fund to help London's poorest and has urged others to follow his example.'

Let me tell you about a very good book that came out in 2008. It is by the BBC's Business Editor, Robert Peston, and it's called *Who Runs Britain? . . . and who's to blame for the economic mess we're in.*[1] Read it, it's very interesting.

Robert Peston grew up believing that 'Britain would become a more equal place'. Then along came Mrs Thatcher and 'Here's the oddity: I gradually came to see that much of what Margaret Thatcher did was necessary'. So Robert now believes that 'we should all cheer when a chief executive pockets millions . . .'

But if it's not spoiling the ending for you Robert does end up, all be it quite politely, blaming the rich for the mess we are in. 'The private sector has been in rude health, especially in London and the south-east. But it is no longer quite so clear that the kind of society being created passes the fairness or the efficiency test.'

Robert deeply regrets that the super-rich fail to see that they should pay something toward the country's upkeep. Even the 'widely respected' Sir Richard Branson keeps the tax man away from his juicy profits by keeping his business empire off-shore.

Robert Peston is forced to conclude by page 343 that 'the argument that the activities of hedge funds and private equity are somehow greatly to the benefit of the vast majority of us is for the birds'.

His remedy lies in the hope that the billionaire class will pay the same proportion of their wealth into the kitty as the rest of us. If that doesn't happen he reckons that in ten years 'Public services would be creaking for lack of resources, as the burden of tax fell on a dwindling number of private-sector employees whose skills weren't quite rare enough or valuable enough to take them into the top league of globally mobile earners.' The rich of course wouldn't have to worry because they would have their own pampered global community and Robert says, 'In that world, elected politicians would seem less and less relevant to the daily lives of the majority.' So then the super-rich would be the rulers. It is a very dangerous situation and Robert's final words are 'We ignore the seemingly unstoppable rise of the super-rich at our peril.'

Let us be grateful then for men like Sir Philip Green showing us that there is another way. The rich are not a lost cause. £100,000 is serious money.

Robert Peston seems to know Sir Philip Green quite well. In fact he dedicates a whole chapter to him. According to Robert, Sir Philip bought a retail organisation called Arcadia in 2002 for a few million and then in 2005 he received a dividend of £1200 million. Actually his generosity got the better of him and he gave every penny to his wife. This was quite handy because she had just become a resident of Monaco. That happy chance meant that £300 million was saved in tax.

You and I might think that was quite clever of Sir Phil but Robert points out that £300 million would have been enough money to build ten state secondary schools capable of educating 13,000 teenagers in total. But of course education is one of the areas that savings have got to be made in; so just as well Sir Phil didn't let the Government get its hands on it. And now he has given £100,000 to the *Evening Standard* and Chris Blackhurst is crowing with delight.

As City Editor he will have no problem working out what that £100,000 represents as a percentage of £1200 million. I just checked my bank balance which is in rude good health at the moment and I wonder if I contribute one twelve-thousandth of it to the fund can he dedicate half a page of the *Evening Standard* to me for my philanthropy. I'm sitting on £1200 at the moment so just let me know where you would like me to send my 10p. As they say it is the thought that counts.

With so much money in one man's pocket you could

be forgiven for believing that there could not possibly be a problem on the planet with anyone lacking enough money to support a dignified life. Surprisingly this is not the case.

Over fifty years ago a Harvard economist, J. K. Galbraith, wrote that, 'Poverty in the US is remarkable'.[2] The fact that poverty still existed in the affluent society of 1950s America he thought was 'remarkable'. He said that our enormous productive power had finally solved man's age-old problem of how to produce enough goods to satisfy our needs and so quite obviously there was no reason for anybody to be living in poverty.

More than fifty years after Galbraith wrote *The Affluent Society*, US government figures show that 40 million US citizens still live in poverty in the richest society that the world has ever seen.

Our extraordinary species has had 300 years – more than ten generations – of incredible technological advance and yet we have ended up in a world where millions are starving while 400 people control the same amount of wealth as one-third of the entire world population.

No tyrant in history ever ruled over a society that got near to that level of inequality. And what is our reaction? 'Well that's the way it's always been. It's greed.' Greed! That is honestly what most people I have talked to about this say to me. Think about it. Greed brings this situation about. How many greedy people do you know? I mean really, really greedy people; greedy enough to explain how 400 people can control the same wealth as 2 billion people. How many are there? Are these greedy people among the billions living in poverty or are they just among the rich and privileged? The Beckhams, the McCartneys, the Queen, Warren Buffett.

Well okay there is that man Sir Philip Greed. But how does this greed operate to achieve this extraordinary state of affairs?

We are so used to what we see around us that we accept the weirdest of situations. It is another reason that societies in the past have been unable to break out of their suicidal systems. The people on Easter Island couldn't see the wood for the trees as they felled the giant timber. 'What are you saying? Stop chopping down the trees? Are you mad? This is life. This is what we do.'

Imagine some voyagers from a distant galaxy happening on our little planet just as European sailors happened on Easter Island. What would they make of it? Surely they would be baffled.

Ten generations of a technological advance so huge that some scientists believe that it is the equivalent of every person on the planet having thousands of slaves working to supply their needs. A technology capable of providing for everybody's needs with plenty left over and yet millions lack even the basics of life. Poverty that exists on a vast scale while the best and the brightest expend more energy explaining and justifying the situation than they do in rectifying it.

Would these space travellers be able to round up enough greedy people to explain how we have got to this ridiculous situation? Or would they point out some simple but basic flaw in our operating system, which we totally ignore and say this is what you need to change, chaps. This is the cause of your problem.

Well given a system where all money is loaned by the banker to everybody else how can it possibly be any different? You may say, but I never borrow from the bank, I don't borrow a penny. I believe in living within my means. I live off my wages and if everybody else would do the same we wouldn't have this problem. Your wages were paid by the company you work for and that company has a bank loan. The money in the world is in the world because somebody went to a bank and the bank created a bank loan for them.

You may resist this idea that all money is created by banks as debt. It may seem too strange to be believable but it is true. Check it out.

But, to convince you that I am right, just try this. What is the UK's national debt?

The UK's national debt is £1 trillion which is a lot of borrowing. And Ed Merrison on Sky News Online on 14 July 2010 said that in fact a truer figure for total government debt is close to £5 trillion. Total personal debt and business debt in this country comes to another £2 trillion. So we either owe £3 trillion or £7 trillion. £3 trillion is twice what we produce in total as a country in a whole year. So now ask yourself where that money was found. It came from the banks as they created it out of thin air. If you disagree please email and tell me where this money was discovered. Which quarry, beach or forest was it taken from?

The present coalition government of the UK is pledged to fight waste. They claim that they will make drastic spending cuts of £20 billion or £60 billion or who knows even £100 billion. The government debt is at the very least £1000 billion and I suspect that actually Ed Merrison is much nearer the mark at £5000 billion. We are buried under debt and yet, with all that money borrowed, huge numbers of people are living in poverty.

Hard-working well-educated people with ten generations of hard-working parents going back to the moment Newcomen first fired up his steam engine. Ten generations of parents who have lived through a period of unprecedented technological innovation. Ten generations of people who have lived in a country that has been amongst the very richest countries that the world has ever seen. And what is the result of all this effort? What is the reward?

We have people living in fear of losing their homes, or worrying about whether their children will be able to have a decent education without ending up with a backpack of debt to start them on their life's journey and we have widespread poverty even in our capital city.

Across our planet of abundance there are massive concentrations of wealth in the midst of huge areas of abject poverty. You can find a million of the poorest on the planet with less wealth between them than a single individual cocooned in a luxury never before witnessed. A gap between rich and poor not matched even closely at any time in human history and still we dare not face the truth.

We will struggle for any explanation we can. It's greed or a lack of charity. We just need one more fundraiser, one more Band Aid concert. We cannot simply admit that the loaning of money into the economy is the cause of poverty. It is impossible for the system to operate in any other way if money is loaned into the economy instead of simply being put into the economy and left to circulate.

But don't take my word for it.

Robert H. Hemphill, Credit Manager of the Federal Reserve Bank of Atlanta, wrote this in 1934:

> We are completely dependent on the commercial banks. Someone has to borrow every dollar we have in circulation, cash or credit. If the banks create ample

synthetic money we are prosperous; if not we starve. We are absolutely without a permanent money system. When one gets a complete grasp of the picture, the tragic absurdity of our hopeless position is almost incredible, but there it is. It is the most important subject intelligent persons can investigate and reflect upon.[3]

10

War: What is it Good For?

Without doubt the country that has used the most firepower in the past hundred or so years is the USA. Supporters of America's wars have argued that America has a manifest destiny to spread freedom and this is why it has fought so many wars.

The Professor of the Science of Government at Harvard, Samuel Huntington says that the identity of the United States is 'defined by a set of universal political and economic values' and they are 'liberty, democracy, equality, private property, and markets.'[1] It is this special identity that makes the United States different from all the great powers that went before them. The United States does not, it is argued, go to war for the same selfish interests that previous mighty powers went to war. The USA is different. There is even a phrase used to describe this difference. It is 'American exceptionalism'.

Samuel Huntington is a professor of the science of government so out of my league, but here is a man who knew quite a bit about war. He was America's most decorated soldier and his name was Major General Smedley Darlington Butler. He ran away and joined the US Marine Corps when he was 16. He won his first Congressional Medal of Honour when he was 19. He rose from private to Major General and won more medals than any other US soldier. He was a focused, efficient commander and his men would follow him anywhere.

In 1932 he retired and went to the US Senate where he made a speech. This is some of what he said.

> War is just a racket. A racket is best described, I believe, as something that is not what it seems to the majority of people. Only a small inside group knows what it is about. It is conducted for the benefit of the very few at the expense of the masses.
>
> I believe in adequate defense at the coastline and nothing else. If a nation comes over here to fight, then we'll fight. The trouble with America is that when the dollar only earns 6 percent over here, then it gets restless and goes overseas to get 100 percent. Then the flag follows the dollar and the soldiers follow the flag.
>
> I wouldn't go to war again as I have done to protect some lousy investment of the bankers. There are only two things we should fight for. One is the defense of our homes and the other is the Bill of Rights. War for any other reason is simply a racket.
>
> There isn't a trick in the racketeering bag that the military gang is blind to. It has its 'finger men' to point out enemies, its 'muscle men' to destroy enemies, its 'brain men' to plan war preparations, and a 'Big Boss' Super-Nationalistic-Capitalism.
>
> It may seem odd for me, a military man to adopt such a comparison. Truthfulness compels me to. I spent thirty-three years and four months in active military service as a member of this country's most agile military force, the Marine Corps. I served in all commissioned ranks from Second Lieutenant to Major-General. And during that period, I spent most of my time being a high class muscle-man for Big Business, for Wall Street and for the Bankers. In short, I was a racketeer, a gangster for capitalism.
>
> I suspected I was just part of a racket at the time. Now

> I am sure of it. Like all the members of the military profession, I never had a thought of my own until I left the service. My mental faculties remained in suspended animation while I obeyed the orders of higher-ups. This is typical with everyone in the military service.
>
> I helped make Mexico, especially Tampico, safe for American oil interests in 1914. I helped make Haiti and Cuba a decent place for the National City Bank boys to collect revenues in. I helped in the raping of half a dozen Central American republics for the benefits of Wall Street. The record of racketeering is long. I helped purify Nicaragua for the international banking house of Brown Brothers in 1909-1912. I brought light to the Dominican Republic for American sugar interests in 1916. In China I helped to see to it that Standard Oil went its way unmolested.
>
> During those years, I had, as the boys in the back room would say, a swell racket. Looking back on it, I feel that I could have given Al Capone a few hints. The best he could do was to operate his racket in three districts. I operated on three continents.[2]

This man is talking about a time when he fought under the US flag starting more than a hundred years ago and continuing up to about 70 years ago. Afghanistan and Iraq are not aberrations brought about because the USA suddenly found itself with a dim rich guy as its President and a gun-toting cowboy as its Vice President.

If a soldier doesn't carry enough weight for you then read what a university professor of equal academic weight to Professor Samuel Huntington wrote in 2010.

> If we turn to the years before 1921, we find the same pattern: Woodrow Wilson's depredations in the Caribbean, the murderous conquest of the Philippines that slaughtered hundreds of thousands, and much else. ... In short, throughout its history, the United States has consistently acted in violation of its ideals.

> But the doctrine that leaders are committed to these ideals is an unchallengeable article of faith.

I am quoting from Professor Noam Chomsky's latest book *Hopes and Prospects.*[3] Let me quote something else from this book because I don't want to give the impression that any of this is about simply knocking America.

> To be sure, U.S. intellectual culture is breaking no new ground. There are two problems with the conventional phrase 'American exceptionalism.' First, to sustain a belief in 'exceptionalism,' one must scrupulously dismiss major parts of what actually happened as the mere 'abuse of reality.' And second, the stance is not peculiarly 'American'; rather it is close to a historic universal among powerful states.

Back through the history of empires from the British through to the Roman, Egyptian and Sumerian, nobody ever said that they invaded somebody else's country for wicked, selfish or greedy reasons. Even 'arch-villains' like the Japanese had honourable intentions; as Emperor Hirohito explained, 'We declared war on America and Britain out of Our sincere desire to ensure Japan's self preservation and the stabilization of East Asia, it being far from Our thought either to infringe upon the sovereignty of other nations or to embark upon territorial aggrandizement.'[4]

Another 'arch-villain' explained the game this way.

> Naturally the common people don't want war; neither in Russia, nor in England, nor in America, nor in Germany. That is understood. But after all, it is the leaders of the country who determine policy, and it's always a simple matter to drag the people along, whether it's a democracy, or a fascist dictatorship, or a

> parliament, or a communist dictatorship. Voice or no voice, the people can always be brought to the bidding of the leaders. That's easy. All you have to do is to tell them they are being attacked, and denounce the pacifists for lack of patriotism and exposing the country to danger. It works the same in any country.

That was Herman Goering.[5]

It may be more palatable to believe that wars happen for noble reasons or possibly because a well intentioned leader made a genuine mistake but the evidence is overwhelming; wars are about power and that means they are about money.

And just to add another voice to the argument. A voice that is not at the end of the spectrum usually associated with Noam Chomsky. This is the voice of our very own Business Secretary in the coalition government, everybody's favourite economist, Dr Vince Cable. This comes from his book *The Storm* published in 2009. He is discussing oil. 'There are reportedly numerous potential fields in Iraq (which, even without invoking conspiracy theories, is undoubtedly one of the reasons for the US presence there). The Saudis argue that there is enormous unexplored potential in the Iraq border area.'[6]

Yes, he is right; oil is 'undoubtedly one of the reasons'. And let's not forget another reason equally locked into money and its close relationship with power. A reason powerfully expounded by Major General Smedley Darlington Butler in his book, *War is a Racket*. The Major General gives numerous examples of how the major corporations have benefited from every war. The profits of these companies sky rocketed each time that the blood of young Americans started to fall on a new battlefield. It is no different today. The profits of Halliburton, Lockheed Martin and numerous others shot up with the launching of the wars against Afghanistan and Iraq.

In a world where everybody is fighting to get their hands on what money there is, a world where there never seems to be quite enough money to go around, then war is always going to be present. And in a world where there has to be more debt than money there will always be a shortage of money.

On page 135 of Niall Ferguson's interesting book *The Ascent of Money* there is a quote by one of the founding fathers of our modern banking and business system, Jan Pieterszoon, a mover and shaker in the Dutch East Indies Company. He said 'We cannot make war without trade, nor trade without war.' Major General Smedley Darlington Butler would support that statement one hundred per cent. War and trade is just the racket he meant when he said a racket is 'not what it seems to the majority of people'.

Of course we are most certainly capable of making trade without war. Just not under the rules of the debt-based money racket that we currently allow.

Just to end this section on war here is what the man who wrote *War and Peace* said about our money system:

> Money is a new form of slavery, and distinguishable from the old simply by the fact that it is impersonal – that there is no human relation between master and slave. (Leo Tolstoy)

11

Cut Here

So the fourth of my compelling reasons for root and branch change was spending cuts. Well-dressed, well-educated men make statements that we accept as making perfect sense. We've overspent. We've borrowed beyond our means. We have to cut back.

It sounds sensible. And of course there could be a big bonus if we remember all those calls to pollute less. It could be a win-win situation. Except that there is a strange twist here. We are not being asked to cut back on the heavily polluting stuff that we do. No, it's education and healthcare and public transport and support services for the vulnerable that we are eager to see trimmed to the bone. We want to cut those areas back because only by doing that can we really get the country moving again by filling more hotel rooms, buying more cars, mobile phones and television sets.

There is nothing wrong with any of those goodies but you have to admit they are not essential to sustaining life. So why do they dictate policy? Well of course we know why: 'It's the economy, stupid'. The question is how we have ended up with this crazy situation and how is it that we buy this daft idea.

Think of your favourite idea for fixing it and see how it would work. How would your favourite solution flip it so that when cuts are needed our intelligent, elegantly dressed leaders beg us to wait another year or two to refit the kitchen, replace the car or upgrade that mobile phone and laptop?

So why do we cut what is important and try to encourage sales of what is relatively trivial?

Remind yourself that money is created by the banks as loans. That money is not loaned to the education service or the health service, it is loaned largely to businesses that pay some of it in wages. Then some of those wages are snatched away to make payments on our national debt that has to be paid to the bankers and some of the taxes find their way into our health service and education. It's not going to be enough of course so the government will have to borrow from the 'money markets' and then the interest payment will increase and a little more of our taxes will be shovelled in to try to keep up the payments to the bankers. The money from the taxes that is left over after the interest on the loans has been paid will be eked out among education, healthcare and public transport.

Our weird way of working with money is again to blame for the fact that when we are told we have been too extravagant with our spending it isn't the luxuries and trivia that are earmarked for pruning, it is the essentials of our human society.

Clever members of our society are lining up to berate us for being too profligate. We are told that our last government borrowed to bribe us with hospitals and nurses and doctors. They spent money extravagantly on education. They spent recklessly on transport infrastructure. And all of that spending was with borrowed money.

After such a telling off what can we ordinary people do but obediently fall into line. Those of us on low pay point to our neighbours as we seethe about the many babies they have, the flat screen TV they bought, the holiday they have just had; and all of it paid for with ridiculous benefits paid by the state. End the benefit culture, we demand.

We snarl at the few groups of workers who unite to protect their income and working conditions. We call them greedy London Underground workers and selfish firemen.

We are pleased when our Bank of England Governor goes to the unions and tells them not to jeopardise our economic recovery by opposing the essential cuts that our coalition government is sensibly implementing. When some troublemaker points to the near £300,000 a year salary and £5 million pension pot enjoyed by the Governor we cry foul. He is different. He does a really important job persuading overseas money markets to bail us out by buying our bonds. Yes, overseas investors are very edgy at the moment. They are getting nervous that our credit rating might be downgraded. They might not trust our bonds any more. They might stop buying them. We could end up like Greece or Ireland.

And it can go on like this forever. We are happy to respond appropriately. It's programmed into us. A hard-wired survival tool – follow the leader. It's safer that way.

Thomas Edison must be turning cartwheels in his grave. He must be screaming fit to bust his lungs: 'If our nation can issue a dollar bond, it can issue a dollar bill. You should be fighting to get your country spending more on healthcare because you could definitely be doing a better job in that area. Do you really think your education service is so brilliant you can start giving it less attention? The money spent is going to pay wages it will circulate in the economy. It is positively beneficial. When all your kids emerge with their heads held high, hearts brimming with high levels of self-esteem; all of them bright eyed and bushy tailed and eager to take an active role in building a better society. Then you can start looking for somewhere else to put your teachers.'

Well, okay, I made some of that up. But he did say: 'It is absurd to say that our country can issue $30,000,000 in bonds and not $30,000,000 in currency. Both are promises to pay; but one promise fattens the usurer, and the other helps the people. ... Why then cannot the people have the benefit of their own gilt-edged credit?'

But we will happily forgo that credit so that we can sink ourselves even deeper in debt as we shovel more of the wealth created with our toil into the pockets of the few.

I said earlier that money is one of our most important creations and should therefore be used carefully. We have handed this incredibly powerful tool to a few people. They have proved to us just how powerful it is time and time again. They have been able to create mountains of money to launch profitable wars. They have filled our cities with skyscrapers that make the statues of Easter Island look positively feeble. They have created a gambling casino that sucks in trillions of dollars in bets on stock and currency movements while most of us huddle in awe outside its doors. And most remarkable of all we regard it as inevitable.

But don't listen to me. Here is the Professor of Economics at the University of Ottawa, Michael Chossudovsky, speaking in 1998.

> Monetary policy is in the hands of private creditors who have the ability to freeze State budgets, paralyse the payments process, thwart regular disbursement of wages to millions of workers and precipitate the collapse of production and social programmes.[1]

12

Stormy Weather

Now the fifth point. A great deal has been written about 'the crash' over the last few years in books like *After The Fall: Saving Capitalism from Wall Street – and Washington* by Nicole Gelinas,[1] a senior fellow at the Manhattan Institute.

Nicole is big on getting the regulations right. And she is in good company after every crash people are shouting about how the regulation has to be improved. It has to be said they usually disagree about what that regulation should be focused on.

But to be fair many of us would agree that most human activities benefit from some kind of regulation and it is certainly worth a lot of effort to get it right. The problem with regulating any area of banking is that the cleverest guys are paid huge sums to do just one thing: to get around the regulations. And these smart operators generally outnumber the regulators by about a thousand to one. And of course there is the obvious fact that regulation has been in the frame for hundreds of years and it hasn't regulated away boom and bust so far.

Two professors of economics from the Universities of Maryland and Harvard – Carmen M. Rheinhart and Kenneth S. Rogoff – take up this very point in their book *This Time Is Different: Eight Centuries of Financial Folly*.[2]

The two professors show us with graphs and charts that man is greedy and foolish and the people who took us into this crash are no different from those who have gone before. And guess what; it will all happen again. So get over it.

Of course we do have our own experts and one of them is Vince Cable, the people's economist. Dr Vince Cable, as already stated, is now our Business Secretary so a man with a great deal of power. He wrote a book which came out in 2009, *The Storm – The World Economic Crisis and What it Means.*

His opinions are taken very seriously, not least because of his perceived integrity. Jeff Prestridge of *The Mail on Sunday* said he is 'everything a politician should be and everything most politicians are not.' And, perhaps from a different part of the political spectrum, Rory Bremner says that he 'gives politics a good name'.

Because of his status and the clarity of his writing and because he is writing at length about the area that really matters – 'the economy, stupid' – I am going to review his book in more detail. If you believe I misrepresent his book in any way please contact me by email and I will take prompt action.

This is how Vince Cable begins his analysis of the recent economic crisis:

> This conjuncture of extreme events and an increasingly hostile political environment has been described as a 'perfect storm'. This short book tries to describe how that storm originated and where it might lead.
>
> Economic storms, like those in nature, come and go. They cannot be abolished. But, as with hurricanes and typhoons, they can be anticipated and planned for and a well-coordinated emergency response, involving international cooperation, can mitigate the misery. They also test

> out the underlying seaworthiness of the vessels of state. The fleet has been plying a gentle swell for some years and making impressive progress. But big waves are already exposing weaknesses. SS Britannica, said to be unsinkable, has sprung a leak, and the vast supertanker USA is listing badly. Passengers and crews are starting to panic and have noticed that most of the life rafts are reserved for those in First Class. How many ships will finally make it back to port in good order after the storm is in doubt.

This is vivid imagery; but there is a difficulty here. The Doctor says, 'Economic storms, like those in nature, come and go. They cannot be abolished.' He is absolutely right about nature's storms. I've sailed a small boat through a few storms at sea and if there was any way of abolishing a storm I would know about it. But the problem is that real storms are created by Mother Nature. They quite definitely will never (I do hope I am right here) obey the will of man. Economic storms though follow the rules of the economy. The economy, however weird it is, is entirely man-made. To repeat, we cannot control real storms because we can never control those natural forces. Economic storms are man-made. Man, not nature, created the system that generates economic storms.

Vince Cable may well be right that they can never be abolished but I want to know why. It is not enough to say, 'Well of course they can't! Think about it! You can't abolish hurricanes and typhoons can you? So there you are; you can't abolish economic storms.' No, it is not necessarily so because hurricanes and typhoons are born of nature while our economic storms are one hundred per cent man-made.

We have constructed our economic system and there has to be a fundamental flaw in the engineering if it throws a wobbly every nineteen years and all that the clever people can do is shake their heads helplessly and wait for two or three years while it pulls itself together.

I mean, yes it's big and, sure, you get so many people involved, and things are bound to go wrong. But the legal system is big and involves a lot of people and goodness knows it can go wrong. Don't get me wrong, I'm not expecting the economy or the law to run absolutely perfectly. I've read John Gray's *Black Mass*. I know such a concept is in the realms of Utopian fantasy.

In the law you have occasional miscarriages of justice or the guilty occasionally get away with it – not good. But you don't suddenly find the entire judiciary standing outside the Old Bailey or the Royal Courts of Justice sobbing into their wigs and demanding a government bailout. The whole system never crashes. Education has its problems but you don't get our entire school and university population suddenly screaming that a tsunami has hit them and they are hoping they can make it to the nearest port. The health service is incredibly complicated and of course things go wrong but you don't have bodies suddenly piling up outside every hospital and health practice for a period of two years while Ben Goldacre explains to the press that they have been holed below the water line.

Okay, you may think I'm mad. You may say, 'This is the economy, stupid. You way underestimate the forces at play here. Yes, it is man-made but that doesn't make it logical. Yes, it would be nice if we all behaved wisely but that is not the real world. And it is certainly not the real economy. People panic, act out of greed, even lust; believe it or not people even lie. No, it is an irrational world. Get over it. The cycle of boom and bust is inevitable. It is the natural cycle of growth and regeneration. Like the passage of the heavens the economic cycle moves through its timed intervals. Intervals that some have even calculated to a nineteen year cycle.'

You may say, as Dr Cable does, that we have known about excess and panic in financial markets for years. As long ago as the dramatic financial crisis of 1824–6 which John Stuart Mill carefully analysed along with earlier events that had occurred over the preceding 110 years. He describes, all those years ago, how the frenzy of overtrading leads to wild speculation. Mill says, 'The failure of a few great commercial houses occasions the ruin of many of their numerous creditors. A general alarm ensues and an entire stop is put for the time being to all dealings upon credit: many persons are thus deprived of their usual accommodation and are unable to continue their business.'

All that is true. Panic and greed, irrational and even devious behaviour, affect everything we do and always will – so, no Utopia then. But millions of vehicles of all shapes and sizes with plenty of irrational humans in control plough up, down and across our irrational network of roads and guess what – accidents happen, breakdowns occur, trains are overcrowded, buses are late or come in threes – but – the whole infrastructure doesn't seize up on a regular 19-year cycle. We don't have journalists reporting traffic chaos from Land's End to John O'Groats or roundabouts in Basingstoke going into meltdown for three years – although in Basingstoke it can sometimes feel like that's the case.

But I have to admit John Stuart Mill is spot on. It is a great description of boom and bust. All the more amazing because the description is still accurate. But it is just that: a description. It is not an explanation. I can accurately describe the symptoms of adult onset diabetes and we can all recognise it every time we come across it. And, if doctors had left it at that, we would all be saying 'Ah yes you're feeling tired, urinating a lot more. Er, have you blacked out yet? No? Don't worry it'll come. Seen it all before.' Yes we sort of need to know the cause so that then we can take action. When we know the cause we say 'Oh okay, I'll change my diet, do some exercise,' or, 'Oh heck just give me the drugs.'

For Dr Cable the cause is as mysterious as our weather system. And he is not alone. Niall Ferguson says that 'Money amplifies our tendency to overreact, to swing from exuberance when things are going well to deep depression when they go wrong. Booms and busts are products, at root, of our emotional volatility.'

So that's it then. The system is fine. It is not even worth looking at it. We can't change it. It is what we are given to work with. The problem doesn't lie in the system. The problem lies in our frailties because 'money amplifies our tendency to overreact'.

Who is he talking about? There are more than six billion of us. I would guess that at least five and half billion don't have anything like the kind of money that would enable them to overreact. So Niall condemns everybody to the pain of boom and bust and spreads the blame across 'our emotional volatility'. The problem is the fault of our unscientific 'emotional volatility', and, just like Dr Cable, he does not believe that economic storms can be stopped.

When it comes to economics it stands alone among the sciences. In all other areas scientists rigorously seek the cause of any effect. They adopt the principle of Ockham's razor and will always work to find the simplest explanation of a given problem. Economists don't do this and for this reason economics has become a popular subject for jokes. Economists lose other mortals with their enormously complex formulas that try to explain our economic system and at the end of it all they will still contradict each other.

In exactly the same way astrology went to extraordinary mathematical lengths to find a working model of our planetary system. They too contradicted each other and were the butt of jokes. Then somebody said let's see what happens if we try our calculations based on the assumption that the sun and not the earth is at the centre of it all.

All that aside though The Storm does give an excellent description of how the crash began which is well worth examining because it certainly illuminates the received wisdom about the workings of our economy.

13

Beware of Falling Rock

Vince Cable says that when the Conservative government deregulated the financial markets Northern Rock was one of the building societies that decided to change its way of working. It became a commercial bank. In 2001 Mr Adam Applegarth arrived. He thought the Rock wasn't being creative enough. Only lending 90% or 95% of the value of the property was for fuddy-duddies. Why not lend 125%? At a stroke you have increased your business by 30%.

Others quickly recognised the genius of this simple idea and the smart money began to move to him. And it was a virtuous circle because customers saw the immediate advantage of what Adam was offering. 'He'll only lend you 95K, you come to the Rock and we'll let you have a 130K – it's a no brainer. They'll only give you three times your income; the Rock will give you six times.'

Adam Applegarth comes across as something of a Robin Hood. He's putting himself out to help poor people and that's got to be good. In fact he's better than Robin Hood. Robin Hood **robbed** the rich to help the poor. Adam **gives** to the rich to help the poor. It's a win-win situation. The share price soared, financial advisers saw their commissions rocket, estate agents sold more houses, the value of every homeowner's biggest investment shot up. Celebration and good times rolled over Jewson, Homebase, Nissan and Ford (125% – that leaves a wad of cash left over for the odd 4x4). Everybody was thrilled. Well nearly everybody.

> **Dr Vincent Cable (Twickenham)**: Is not the brutal truth that with investment, exports and manufacturing output stagnating or falling, the growth of the British economy is sustained by consumer spending pinned against record levels of personal debt, which is secured, if at all, against house prices that the Bank of England describes as well above equilibrium level?
> **Mr Brown:** The Hon. Gentleman has been writing articles in the newspapers, as reflected in his contribution, that spread alarm, without substance, about the state of the British economy ...[1]

And the points go to the Doctor.

Adam Applegarth – now wasn't he just doing what the game demanded? If he hadn't done it wouldn't somebody else have come along and said to the board, the shareholders, the customers, the homeowners, the retail industry (in fact 'the economy, stupid'), 'Come on you know you want it and I am just the man to give it to you.'

So I take a slightly different position to the Doctor on this one. When he tells me that Adam 'acquired fast cars and a castle' I say fair enough; who wouldn't if they were given the opportunity? Let us be candid, a man with the brass balls to get an entire economy steaming across the ocean at 28 knots is not going to be the kind of fellow to spend his leisure hours dangling a line into his local canal while sipping from a Thermos and nibbling on a Spam sandwich.

And what would the *News of the World* write about if

it didn't have interesting people like Adam Applegarth around? Apparently 'a mistress' ('a' mistress? Did he have more than one? Sorry but he has risen even higher in my estimation.) 'was rewarded with five mortgages and a property empire'.

Come on, he deserved it. In just three years he doubled Northern Rock's share of the mortgage market. Yes, I said doubled, that's what the Doctor said. Think about that. How much would that be in pound notes? A lot. This isn't the Pound Shop that Adam Applegarth is sprinkling his fairy dust on to. His sales are one hundred, or two hundred grand a pop. And he doubled the business in three years.

Should he now feel ashamed? Well not if you believe in the free market. Especially that strand that says there is no such thing as society. Or that the business of business is business. Or that money is the bottom line. What else should the man have done?

The Doctor isn't so easy on him. 'Northern Rock forms a central part of my story because it was the small hinge on which the British economy swung.' The Doctor shows clearly that in blowing up the housing bubble which finally burst in 2007, 'Easy credit was the key. Competition among mortgage lenders produced a bewildering variety of mortgage products – 15,600 in July 2007.'

15,600 different ways to buy a house!? Adam Applegarth was not steering this ship single-handed. No indeed! 'Northern Rock was not the only bank willing to lend 100 per cent or more of the value of a property and five or six times the borrower's income.' And Dr Cable has this to say as well: 'Mainly because of mortgages, but also because of personal borrowing, average household debt has risen to 160 per cent of income, double the 1997 level – the highest of any developed country, and the highest in British economic history.'

Unbelievable; but it gets even worse because, 'The UK is merely one, modest, part of the global economy: barely 2 per cent of it. The collapse of confidence in financial markets and in what were, until recently, seen as stable institutions is a much wider phenomenon.' In the US 'a veritable army of Adam Applegarths pumped out enormous numbers of mortgages, often aimed at poorer families or those with a poor credit history. So called "ninja" loans – to people with no incomes, no jobs and no assets – look in retrospect to have been criminally irresponsible. But at the time it seemed a worthy idea . . .'

Now what seems criminal to me is not lending money to people to buy a home. No what I find criminal is that there should be people – lots of them – living in the twenty-first century, in the richest society that the world has ever seen, who have 'no incomes, no job and no assets'. And totally criminal that predators should be freely able to swim among them and saddle them with debt so that they not only have nothing, they have less than nothing. Such treatment should be alerting some members of the UN to consider a policy of regime change.

But I'm sure there would be people who would say to me that these people didn't have to take on those loans which then triggered the collapse of the world's sophisticated financial system. These poor people, sub-prime people, took on $1.3 trillion in mortgages.

$1.3 trillion! A trillion is a million multiplied by a million. John Browne from Europacific Capital speaking on Fox News said that, counting one dollar a second it would take 41,000 years to count one trillion dollars. Where did these Adam Applegarths in America find that kind of money?

I suppose I know the answer, because Vince says in his book that our Adam Applegarth got his 'mortgage lending funds from the wholesale markets'. So I guess that's where the American Applegarths got theirs.

I tried to Google 'How many wheelbarrows would it take to carry off $1.3 trillion?' I did find out that a pallet load of $100 bills adds up to $100 million. So $1 billion would be ten pallet loads, so a $ trillion would need to be stacked neatly onto 10,000 pallets. Having moved pallet loads of sacks of sand I would say you would need 10 wheelbarrows per pallet so that is 100,000 wheelbarrow loads to pick this money up at the wholesale market. And that would be cheating because this money was for poor people and they wouldn't know what a $100 bill looked like so it would have to be in $1 dollar bills really, which means you would need 10 million wheelbarrows.

So that's another question: where do I find these wholesale markets, because I just happen to have a couple of wheelbarrows just waiting for something to put in them.

Now that I think about it I am reminded of another question that Richard Dawkins put in my head when he was talking about God. Something like – if the answer to 'How did everything get here?' is that God made it, then the next question, is who made God? So I need to ask Dr Cable where did these people who run 'the wholesale markets' find these 10 million money-filled wheelbarrows? I mean where did the money market chaps find this wad of cash? Because it must come from somewhere, it doesn't just grow on trees, does it? Well actually if you are a banker apparently that is exactly where it does come from.

Anyway these poor people were on 'teaser' rates, low introductory interest rates, and of course all good things must come to an end. When rates suddenly rose these poor people couldn't make their repayments and a lot of them didn't even wait to be repossessed, they just handed in the keys and the bubble burst. House prices fell by 25 per cent.

> The number of repossessions has been variously estimated at 2 Million on the conservative side to as many as 6.5 million by Credit Suisse – as many as one in ten mortgages.

I get the feeling we might have the wrong people doing the counting. The Doctor is telling us they don't know if it's 2 million or 6.5 million who lost their homes. This isn't a miscount of ballcocks at B&Q – these are people. I would like to know how many minutes the Doctor spent visualising bailiffs walking up to his house and explaining that unfortunately he is going to have to relocate – along with his weeping wife, three children and the Labrador. It would devastate me.

> Whilst this story was distressing for those American families, it is not immediately obvious why their problems should have reverberated around the world. [That is economist-speak for empathy.] To understand this, I need to explain how the US mortgage market works and how its risks are transmitted to wider financial markets. The total US mortgage market was worth roughly $12 trillion in July 2008. [Well as it turns out] 'sub-prime losses simply do not justify the collapse of confidence that has occurred. ... In fact, when the IMF made its estimate of total US financial sector losses in its Global Financial Stability Report, it

> estimated that of $1.4 trillion ($1400 billion), only around $150 billion could be traced to mortgages, and only a fraction of that to sub-prime mortgages.

So the poor folks were not to blame after all? No they very definitely were not. The real problem was the 'non traditional lending outside the banking system': This is what is known as securitisation. 'Through securitisation, loans once held on the books of banks were repackaged and sold.'

Loans were sold? It is difficult to get your head around, but these institutions, or people, or whatever, were selling debts; this was their business! In fact it was worse 'institutions borrowed money in order to buy debt, [I'm with it so far] which was the security for the borrowing, and the money they borrowed was in turn borrowed, sometimes through several institutions.' No I'm lost, sorry. But there's more! 'In addition, debt default could be insured against, but the insurers depended in turn on borrowed capital.' I think I am slowly beginning to see how things could go wrong with this working model.

'Derivatives markets also made it possible to hedge (or speculate) against risk of default.' If I speculate don't I expose myself to risk? So derivatives allow me to take a risk against a risk? 'The credit default swap market, for example, which grew on the back of the growth of these debt instruments, achieved a notional value of over $60 trillion.'

Wait! 60!? Trillion?? 'This, in turn, represented about one tenth of the overall size of derivatives markets, which Warren Buffett warned us was the H-bomb to follow the sub-prime A-bomb.'

The derivatives market is $600 trillion! $600 trillion is $600,000 billion. $600,000 billion is $100,000 in the back pocket of every man woman and child on planet earth – and that's just one market place!

A $600 trillion market in anything should surely alert us to things being very, very out of whack. And it is no use saying, 'Well of course it's not real. You have to understand the way these things work. It's notional.' If that's the case why is Warren Buffett delivering dire warnings?

It is often said that in telling a story or revealing any truth it is better to show not tell. Think about what Dr Cable has just shown us. The incredible complexity of 'securitisation ... loans repackaged and sold ... money borrowed was in turn borrowed, sometimes through several institutions ... debt default could be insured against, but the insurers depended in turn on borrowed capital ... Derivatives markets made it possible to hedge (or speculate) against risk of default ... debt instruments, achieved a notional value of over $60 trillion.'

This is a staggering complexity that even the Doctor's impeccable writing style can not make clear for most of us mortals. And he shows the equally staggering simplicity with which these mountains of money were created. $60 trillion! And that is only one tenth of the money that they managed to create.

Dr Vince Cable graphically shows us the truth that Professor John Kenneth Galbraith tells us:

> 'The study of money, above all other fields in economics, is one in which complexity is used to disguise truth or to evade truth, not to reveal it. The process by which banks create money is so simple that the mind is repelled.

Vince Cable's book is very good. There is an excellent description of how the US government and the Federal

Reserve poured in taxpayers' money to save some of these financial institutions. He describes the effect that the oil market has on the world economy and especially on food prices. He talks about the huge impact China has had on the world economy and what it will be in the future. And indeed the impact other emerging markets, especially India, are going to have. It is very interesting but I am now going to concentrate on his summary and suggested remedy for this crisis and how he would prevent future crises.

Vince Cable says that: 'A key step was to recognize ... that banks should have whatever liquidity is necessary from the central bank (albeit at a penalty rate and secured against sound collateral)'. He favours the UK government efforts over the US government plans, some of which were 'badly conceived and politically unpopular'.

The UK government's method was to inject money into the banks by way of a partial nationalization – I can hear Karl Marx singing, or maybe not. 'The state preference shares enjoy a 12 per cent interest for taxpayers who receive no dividends on the ordinary shares.' No dividends? Now why would that be? Organise a devaluation of the currency of say 12 per cent over the next year or two and the banks get the money for free. But they wouldn't do that would they? I suppose it depends on how many fast cars and mistresses they have to keep on the road.

Just so you know, the money the UK Government gave to the banks was climbing past £850 billion by the spring of 2010. Of course as you know the UK Government doesn't have any money. That's right, it has that trillion pound (£1000,000,000,000) debt. So where on earth did it get that other £850 billion? Yes, the money markets. Times are bad but the money markets discovered another 5 million or so badly parked wheelbarrows heaped with money. They've handed the money to our Government because the money markets are generous and consider our Government a good risk because our Government has a limitless source of future wealth. Us.

'Other than the interest rate, the main attraction for the taxpayer is that the banks have agreed to a (rather vague) undertaking to maintain lending and to restrain bonus payments.' Those words in brackets don't make you feel very secure do they? And 'maintain lending'? Isn't that what they usually do? They are banks after all, and what does 'restrain bonuses' mean? Still Vince thinks it's as good a deal as we can get.

He also thinks it's sensible to cut interest rates and to create a 'fiscal stimulus by, temporarily, running a larger budget deficit than would normally occur even in a downturn'. He acknowledges that some would disagree with these prescriptions. 'There are those who worry that governments are acting precipitately, however, and risk creating even bigger problems in the future.'

Alternatively, he suggests that you could nationalise failing institutions, replace the management, and then sell them on in improved conditions. In which case I wonder why he doesn't think they should just remain nationalised? But he doesn't. He says that, 'far from temporary nationalisation being a step towards socialism, it is an essential tool for managing a market economy and maintaining its disciplines in a financial crisis'.

Vince did suggest to Gordon Brown in the House of Commons, as quoted, that there was cause for concern, and he says in his book that, 'Perhaps someone in the Bank of England or the Treasury should have stopped to think about "what if" scenarios, such as the risk of a small but ambitious bank behaving recklessly and putting its depositors at risk. But no one did. Until Northern Rock.'

Of course Vince has admitted that Northern Rock wasn't alone. And who in government is going to stand up when the entire market is booming and shout 'Hey! Slow down.' And in any case what is the Doctor saying should have been done to halt the stampede?

So we are left knowing that there 'are a lot of questions about lending practices, particularly in respect of sub-prime loans, and on mis-selling and the aggressive promotion of debt. But this is essentially an issue of the regulation of lending practices. Few would advocate large scale debt waivers, since the moral hazard in encouraging future excessive borrowing is obvious.'

'Banking is an Alice in Wonderland world where over and over again recklessness has led to calls for help.' And even this time round, when we should have known better, greed and stupidity have led us down the same path. But a 'system that allows banks and other institutions to make profits and fat salaries from questionable and foolish practices, while the public picks up the bill, should simply be unacceptable'.

I think we would all agree with that and Dr Cable promises to offer suggestions for reforms in a later chapter so that is where I am going.

14

The Doctor's Prescription

The final chapter of *The Storm* is titled 'The Future: A Road Map'. What the Doctor's remedy boils down to is that the Government should 'borrow and spend in order to maintain the level of activity of the economy'. Borrow money and give it to consumers by cutting the amount they pay in taxes. Always a popular choice I would think. Then borrow more money and give it to businesses by paying for new social housing and build new roads and railways.

Any other suggestions? Yes, encourage new lending through state guarantees or, if necessary, bypass the banks and lend directly to businesses. (Don't get too excited – he's talking about big businesses here.)

So our problem was caused by too many people taking on too much debt and the answer to the problem is for more people to take on some more debt. I am not saying it won't work but it does appear to be somewhat, what's that word, counterintuitive. And the reason to take on more debt is so people can make more goods so the economy will be happy with us.

Okay, there are two other things we can do according to the Doctor. The first is get the regulations right. There are others who say that this is a good idea; unfortunately they often have completely opposite views on just what these regulations should be. And it has to be said it hasn't worked in the past – these crashes do have a habit of repeating themselves; inevitably, Dr Cable tells us.

The second point that he makes is that we must keep talking to our international partners especially to keep trade going. Consider this statement though and ask yourself if it makes you feel confident that cutting interest rates, borrowing more money, working on the regulations and continuing trade talks is going to resolve the problem.

> We are still left with a series of interconnected markets, which were valued by the Bank of International Settlement in 2007 at $516 trillion, thirty-five times the size of the US economy in GDP terms, ten times the total size of the world economy, five times the size of all the world's stock and bond markets, and seven times the size of all the world's property markets.

For the Doctor though 'The problem remains how to prevent a rogue 1–2 per cent of the market going wrong . . . .' I am not sure that it's the 1–2 per cent that should concern us. I think we need to pull the curtain back from the other 98 or 99 per cent.

I am not as convinced as Dr Cable that fiddling with interest rates will make it all right. Nor do I believe that government spending, personal spending, business spending is a sustainable solution. Especially when I remember that the Doctor said that spending will be made up of government debt, our government, so ultimately we the taxpayers are responsible for repaying it.

No, that is not an answer and nor is regulation. As already pointed out financial experts are highly paid in order that they will find ways around any government regulation. Neither do I think more trade is the answer. So I do have to agree with Vince Cable when he says that economic storms cannot be abolished. He is absolutely right if the above solutions are the only ones available.

Let me quote from the book's final paragraph.

> Last but not least, the development agenda – to eliminate hunger, poverty and disease – for which the World Bank is the lead agency, has to remain central, for both economic and moral reasons. Looking back on the events of the last few months, what is striking is the alacrity with which the USA and the EU have managed to mobilize $3 trillion (and rising) in capital and guarantees for failed banks, having failed to mobilize $300 million to help fight hunger in the midst of a food supply crisis earlier in the year. Such narcissistic self absorption and twisted priorities do not bode well . . .

I get a bit lost with all those millions and trillions but it sounds like the Doctor is saying that some rich people just couldn't locate $300 million to feed people who were starving to death but they suddenly discovered 10,000 times that sum when the banks got confused over their accounting systems.

I wonder if many people would describe that as 'narcissistic self absorption and twisted priorities'. I will leave it to you to consider how you would describe such behaviour.

The Doctor's suggestion to help the 30 million poor people – the world's sub-prime people – who will watch their children die ever so slowly over the coming months is to put in place 'a structure of global governance in which the main emerging economies have parity' because he believes this 'would do something to redress the balance'.

The main emerging economies can sort out poverty. India and China need to be given an equal voice at the trade talks and that will redress the balance. Let India and China bring their expertise to the table. They should have a lot of expertise given the number of poor people they have. Maybe they can advise the rich countries like the US and the UK on the best way to 'eliminate hunger, poverty and disease'. Issues which Dr Cable says, on the last page of his book, must 'remain central, for both economic and moral reasons'. And for which, he tells us, 'the World Bank is the lead agency'.

I'm not sure what the poor will make of the Doctor's treatment plan but I can't say that it fills me with confidence.

15

The Bottom Line

So is there a different way of looking at this problem, beyond tweaking interest rates and regulations and hoping that commission driven mortgage salesmen and hedge fund managers will suddenly stop being greedy?

Well the Earl of Caithness certainly thinks so.

> [O]ur whole monetary system is dishonest, as it is debt-based ... We did not vote for it. It grew upon us gradually but markedly since 1971 when the commodity-based system was abandoned. (The Earl of Caithness, in a speech to the House of Lords, 1997)

Going back to basics, the system operates on money. Money is put into the system in the form of loans which then have to be repaid. Earlier on we saw how this makes for a struggle by people in the system to get their hands on the money so that they can buy the things they need and make repayments to the bank. It is mathematically impossible but it is a brutal motivator.

There always have to be many more losers than winners. Even if every human on the planet was as dynamic as Richard Branson, as clever as Vince Cable and as ruthless as Philip Green, losers would still outnumber winners by exactly the same proportion as they do right now because it is a simple question of mathematics. Among those losers of course is the environment.

Now to really show how the hurricane blows into the economic weather system we have to make it a little more complicated. I mentioned interest charges before but didn't fully describe the effect they have. Going back to our hundred people on the island we imagine the banker giving them all £100,000 at 8% compound interest. Not only are they fighting each other to get enough money to live on and make repayments, but they have to find enough money to cover the 8% interest.

Think carefully about this. Where can this 8% come from? The banker gave each person £100,000 but he wants more than £100,000 back from each person. In fact they will have to find another £100,000 on top of the £100,000 that they borrowed if they make repayments stretch over 20 years.

But the banker didn't put the extra money into the system. The extra money to pay the interest on the loan can only come from one place. It has to come from the banker. He will have to make further loans. Further loans made with interest on them.

As we start out on the business cycle loans will be made. A big company will see a major investment opportunity and borrow from the bank. Sub contractors will see that there is a market for them in supplying this big company and they will go to the bank. Workers will see a period of prosperity and take out that loan on a new house or a car. All of these loans mean of course that a lot of new money is now circulating in the country.

The problem is though that all this money has been put into the country as a loan from the banks and it has to be paid back to them. The money is being paid back in repayments to the bank so the money is leaving the economy. The banks will obviously want to get this money back into the system so there will be enough money to pay back the earlier loans including the interest. Great deals on mortgages will flow, with or without an Adam Applegarth, mail outs will tempt us with new credit cards. Banks will offer 'amazing' deals on new car and kitchen loans.

The economy now begins to boom. Maximum employment, easy credit, property prices rising meaning more money can be borrowed. And it's essential that it is borrowed because the amount of money now needed to repay the loans out in the economy is enormous. And remember those loans have interest attached to them. But that's all right because we just need to find more people to take on more loans. And the wonder of the system in the boom times is that there are always wheel-barrow loads of money to be hauled out of the banks.

At the peak of the excitement the banks are generating new loans so quickly that the economy is awash with money and people are buying homes as an investment and magically they are increasing in value. Stocks and shares get the same treatment and more loans wash into the system and this time it will run for ever. This time will be different. This time a trillion pounds will pump its way into the economy in the form of loans.

The trillion pounds was created by the banks but over the lifetime of those loans a trillion pounds worth of interest will be added. A trillion pounds of interest that the banks insist their borrowers pay to them. Money that does not exist because it was never created. And that is why there is always more debt than money created.

And just when the balloon is stretched to its maximum the smart men, the big players begin to dump shares and property and then sit back and watch rout begin. And, when the havoc is total, slowly and quietly, the businesses get bought out at a penny in the pound and so now even fewer people are in control of the wealth of the world.

That's how it works. Money created with interest so more money has to be created to pay the interest on the earlier loans and it builds again. And that's the business cycle.

The problems we face are not caused by greed, or poor regulation or the wrong interest rates or even wasteful governments. It is simply our ignorance of how our money system works. Or as President John Adams put it: 'All the perplexities, confusion and distresses ... arise not from defects in the constitution or confederation, nor from want of honor or virtue, as much from downright ignorance of the nature of coin, credit, and circulation.'

There it is stated clearly 200 years ago. President Adams does not resort to colourful meteorological metaphors about 'economic storms' that come and go, which like 'hurricanes and typhoons ... cannot be abolished'.

Nor does he search into the world of psychology for truths such as 'money amplifies our tendency to overreact' which shows us that 'Booms and busts are products, at root, of our emotional volatility.'

Like his friend Thomas Jefferson, he would have been somewhat baffled by the analysis of Dr Cable and Professor Ferguson. Adams and Jefferson were practical philosophers.

I like their explanation. I certainly prefer to see us as intelligent beings able to make sensible choices. I am especially keen to see us making sensible choices when those sensible choices are in an area that has such an enormous impact on every person on the planet. And indeed that have such an enormous impact on the planet.

I should also say once again that this is not in the area of rocket science. This is a science that is at root about connecting the goods and services that we produce with the people who need them.

President Jefferson and President Adams were in agreement about what was the major cause of their problems and Jefferson put it most succinctly when he said: 'I sincerely believe, with you, that banking institutions are more dangerous than standing armies.'[1]

A more recent commentator is the Belgium professor of economics Bernard Lietaer who wrote:

> I have come to the conclusion that greed and fear of scarcity are in fact being continuously created and amplified as a direct result of the kind of money we are using. For example, we can produce more than enough food to feed everybody, and there is definitely enough work for everybody in the world, but there is clearly not enough money to pay for it all. The scarcity is in our national currencies. In fact, the job of central banks is to create and maintain that currency scarcity. The direct consequence is that we have to fight with each other in order to survive.

He goes on to say:

> Money is created when banks lend it into existence. When a bank provides you with a $100,000 mortgage, it creates only the principal, which you spend and which then circulates in the economy. The bank expects you to pay back $200,000 over the next

> 20 years, but it doesn't create the second $100,000 – the interest. Instead, the bank sends you out into the tough world to battle against everybody else to bring back the second $100,000.[2]

We need to withdraw the right to create money from banks. They can take it in and store it safely and they can give out loans for which they can charge a fee to cover their costs like any other business. But they cannot be allowed to create the money.

The power to create money must be under full democratic control. Then of course we will not be 'so-called' democracies we will be fully functioning democracies. The rulers of any state are the people who have the power and nobody has real power unless they control the money. If we want true democratic control we must have full democratic power.

16

The Descent of Money

I realize that many of my expert witnesses come from long ago so just to show that what I am saying is supported by modern day experts in the field I am going to Niall Ferguson's book *The Ascent of Money.*[1] Apart from being one of Britain's most renowned historians he is the Laurence A. Tisch Professor of History at Harvard University.

You will remember I have already referred to him. Once to show how much money influences the course of history, a point made clearly by Professor Ferguson in his comment that: 'The Crusades, like the conquests that followed, were as much about overcoming Europe's monetary shortage as about converting heathens to Christianity.' I also quoted him to support Dr Cable's belief that man is pretty much at the mercy of the economic typhoons that strike upon us regularly and inevitably. Niall's take is that 'money amplifies our tendency to overreact, to swing from exuberance when things are going well to deep depression when they go wrong. Booms and busts are products, at root, of our emotional volatility'.

From that it must follow that the only way to eliminate boom and bust is to eliminate emotional volatility. Definitely not something I would want to do. I would be the last person to campaign for the elimination of human volatility. Even to eliminate boom and bust universal lobotomisation is just too drastic.

Fortunately the Professor shows that he doesn't really believe this.

On pages 30 and 31 we learn that:

> Today's electronic money can be moved from our employer, to our bank account, to our favourite retail outlets without ever physically materializing ... The intangible character of most money today is perhaps the best evidence of its true nature. Money is not metal. It is trust inscribed ... And now, it seems, in this electronic age nothing can serve as money too.

So in our electronic age nothing can serve as money. Money out of nothing; which is pretty much what most of the expert witnesses from the past have said. As for instance the previously quoted Graham Towers, Governor of the Bank of Canada: 'Banks create money. That is what they are for . . . The manufacturing process to make money consists of making an entry in a book. That is all . . . Each and every time a Bank makes a loan . . . new Bank credit is created – brand new money.' Brand new money out of nothing.

But there is a difference. The Professor is in total *dis*agreement with US President John Adams when he said that, 'All the perplexities, confusion and distresses . . . arise not from defects in the constitution or confederation, nor from want of honor or virtue, as much ' from downright ignorance of the nature of coin, credit, and circulation.' And with President Jefferson when he said, 'I sincerely believe, with you, that banking institutions are more dangerous than standing armies.'

On pages 64 and 65 Ferguson says:

> [B]anks have evolved since the days of the Medici precisely in order (as the 3rd Lord Rothschild succinctly put it), to 'facilitate the movement of money from point A, where it is, to point B, where it is needed.' Credit and debt, in short, are among the essential building blocks of economic development ... Poverty, by contrast, is seldom directly attributable to the antics of rapacious financiers. It often has more to do with the lack of banks, not their presence.

No he is very definitely not on the same side as Jefferson. And certainly I have to take this opinion seriously. Niall Ferguson started life in Glasgow so apart from his academic prowess he will have seen poverty at close quarters. But also he is no stranger to the other end of our society. His professorship is named for benefactor Laurence A. Tisch. Tisch was a billionaire and CEO of CBS. Henry Kissinger granted access to his private papers to Niall who wrote a biography about the man and indeed he had privileged access to Lord Rothschild's papers when he wrote the biography of the Rothschild family.

Now looking back to the quote from pages 64 and 65 he says that banks 'as the 3rd Lord Rothschild succinctly put it "facilitate the movement of money from point A, where it is, to point B, where it is needed."' Yes, succinctly put.

Two things immediately strike me. If this is what they do should we not be concerned that, with a billion starving through lack of purchasing power, these banks are not perhaps a great deal more effective than their Medici forebears at facilitating the movement of money.

Of course Niall might say that in places on the planet

where poverty is at its most intractable there are very few banks. He does in fact make this very point in some detail when he gets to applauding Peruvian economist, Hernando de Soto.

Of course one area of the planet that has had no shortage of banks for the past three centuries is the City of London. And indeed the mother of all banks still threads her industrious needle in its very heart. The Bank of England was founded to facilitate the movement of money. It and its banking army have moved it with such ruthless efficiency that today 400 people are stuffed to the gills with the stuff while the *Evening Standard* runs its appeal to help the poor who live within spitting distance of the Bank of England. Three hundred years of facilitating the movement of money and they can't even get it to where it is needed in London.

We have had several centuries of bank evolution 'since the days of the Medici' so it does seem time, given the essential part played by banks in our economic development, that poverty, whatever the cause, should have been banished.

My second point is that the statement seems to be a little too succinct. Where is the reference to the banks' other role? That is the creation of money. Perhaps the 3rd Lord Rothschild was too busy doing whatever it is he does to mention that part of the banks' function.

It is on page 122 and 123 that we are told the way things really operate. The Professor tells us that we have had four hundred years of financial bubbles. 'Time and again this process has been associated by skullduggery, as unscrupulous insiders have sought to profit at the expense of naïve neophytes. So familiar is the process that it is possible to distill it into five stages.' He then describes 1) New opportunities opening up. 2) Rising profits lead to rise in share price. 3) First time investors are attracted as well as 'swindlers eager to mulct them of their money.' 4) 'The insiders discern that expected profits cannot possibly justify the now exorbitant price of the shares and begin to take profits by selling.' 5) 'As share prices fall, the outsiders all stampede for the exits, causing the bubble to burst altogether.'

So there it is a brilliant description of how our 'emotional volatility' has got us into trouble. However Niall does confess that 'not everyone is irrational' the 'seasoned speculator' and 'the true insider' are much more likely to get their timing right than 'the naïve first-time investor'.

But then the bombshell in a single sentence on page 123:

> Finally, and most importantly, without easy credit creation a true bubble cannot occur.

Absorb the Professor's unequivocal statement and then remember what he said in his Introduction on page 15: 'Booms and busts are products, at root, of our emotional volatility.'

Without easy credit creation no amount of 'emotional volatility' is going to get a bubble going. At root, Professor, is easy credit and not emotional volatility.

And who supplies the easy credit? Those banks who Lord Rothschild and Professor Ferguson tell us 'succinctly' are merely acting to 'facilitate the movement of money from point A, where it is, to point B, where it is needed'.

Look again at how the Professor describes the financial bubbles of the past four hundred years. You would be forgiven for believing it is the 'skullduggery' of 'unscrupulous insiders' that is to blame. Combined of course with our 'emotional volatility'. The banks desperately and succinctly moving money from A to B are merely helpless bystanders.

Miss that one sentence ('Finally, and most importantly, without easy credit creation a true bubble cannot occur.') and you have missed the real point. The endless cycles of boom and bust can only occur because banks can create credit on a vast scale and reduce credit when the time is right.

Jefferson recognized this 200 years ago when he warned of the danger of allowing private individuals to control a nation's money. He said they had used that power 'to inflate, by deluges of paper, the nominal prices of property, and then to buy up that property at 1s. in the pound'. And he went on to say that:

> This is what has been done, and will be done, unless stayed by the protecting hand of the legislature. The evil has been produced by the error of their sanction of this ruinous machinery of banks; and justice, wisdom, duty, all require that they should interpose and arrest it before the schemes of plunder and spoliation desolate the country.

Two centuries of enormous technological progress and massive wealth creation have elapsed in the United States since he spoke those words. What would his reaction be to the result of all that extraordinary effort?

More recently Congressman Wright Patman who was chairman of the House Committee on Banking and Currency for 40 years said in Congress in 1941:

> When our Federal Government, that has the exclusive power to create money, creates that money and then goes into the open market and borrows it and pays interest for the use of its own money, it occurs to me that that is going too far. I have never yet had anyone

> who could, through the use of logic and reason, justify the Federal Government borrowing the use of its own money ... The Constitution of the United States does not give the banks the power to create money, but now, under our system, we will sell bonds to commercial banks and obtain credit from those banks. I believe the time will come when people will demand that this be changed. I believe the time will come in this country when they will actually blame you and me and everyone else connected with this Congress for sitting idly by and permitting such an idiotic system to continue. I make that statement after years of study.[2]

This is exactly the point that Thomas Edison was making.

The Congressman was saying that the government does create our money. It creates a bond. A bond is simply a promise to pay. Here is my bond, my pledge, my promise signed on behalf of a nation's government that a million pounds will be paid. The banks have faith in that promise to pay. (Because an army of we taxpayers stand ready to work to produce taxes to make good that promise to pay.) So they accept it as being worth a million pounds so they hand over a promise to pay a million pounds. The bankers' promise to pay however comes with an attachment. The government in this transaction is agreeing to pay interest. Interest that will grow as a national debt.

The government's 'bond' is nothing more than a promise to pay. It is only useful if it is trusted as having the value stated. The banks' 'money' is nothing more than a promise to pay. It is only useful if it is trusted as having the value stated. And the banks trust the government's promise to pay – so why shouldn't you? Think of all the interest you will save.

Niall Ferguson's *The Ascent of Money* is a detailed

history of money which as one reviewer on the back cover of the book says is 'breathtakingly clever'. We learn about 'the man who invented the stock market bubble', John Law. Law was born in Edinburgh in 1671. 'According to Law, confidence alone was the basis for public credit; with confidence, banknotes would serve just as well as coins. "I have discovered the secret of the philosopher's stone," he told a friend, "it is to make gold out of paper."'

'According to Law, confidence alone was the basis for public credit'. According to Professor Ferguson as well of course. As he said on page 30: 'What the conquistadors failed to understand is that money is a matter of belief, even faith: belief in the person paying us; belief in the person issuing the money he uses or the institution that honours his cheques or transfers. Money is not metal. It is trust inscribed.'

Today, it appears we are even cleverer than John Law. He performed his magic with the help of paper whereas today Niall tells us that our John Laws don't even need to supply the paper: 'in this electronic age nothing can serve as money too'.

John Law appears quite delirious with the magical simplicity of this philosopher's stone that is our banking system. Niall Ferguson also appears to be in awe of it and the intellectual giants who control it.

Congressman Wright Patman as we know took a different view.

> I believe the time will come in this country when they will actually blame you and me and everyone else connected with this Congress for sitting idly by and permitting such an idiotic system to continue. I make that statement after years of study.

17

Democracy Now

Whether you count yourself a Conservative, Socialist, New Labour, Liberal Democrat, UK Independence, Green or Other, the question I want to ask you is: are you a believer in the idea of democracy?

Democracy is often described as the rule of the people for the people by the people which I imagine most people would go along with. Now whichever party you support can you honestly say that is what we have? No matter how fervent a democrat you are you know there are those who say, 'I don't vote because it really doesn't change anything.' Or even, 'If voting did change anything they would ban it.'

Now you are probably not that cynical but you know that our elected government is not free to bring about a better world for all. It has to work within the real world of power. President Clinton's campaign manager, James Carville, put it most succinctly when he came up with that election winning slogan, 'It's the economy stupid'.

If we want true democracy then we must have the courage to take control of that fundamental source of all power, money.

That's not easy. First people need to understand how our money system operates and not many people do.

Then they need the courage to stand up to the inevitable assault. This system does work incredibly well for a few people, many of whom are extremely persuasive. The ebullient Kenneth Clarke is very persuasive but he is hardly going to be a supporter of the notion that only a fully accountable body should be in charge of the money supply of our nation.

He is a member of parliament and now Lord Chancellor, and a very pleasant personality who is always entertaining on television. But of course he is also a Director of Foreign & Colonial Investment Trust, Member of the Advisory Board of the hedge fund management company Centaurus Capital, Deputy Chairman and a director of British American Tobacco, Chairman of Unichem, and Director of Independent News Media to name a few.

It would obviously be unfair to expect Kenneth Clarke to enter a discussion on the pros and cons of the debt-based monetary system that currently operates. Obviously; since the companies who pay him his director's fees are companies that most certainly are beneficiaries of our present monetary system.

It took a long time for the idea of our sun-centred solar system to be accepted. There are always good reasons for resisting radical change of our core beliefs. And Machiavelli was undoubtedly right when he said:

> There is nothing more difficult to take in hand, more perilous to conduct, or more uncertain in its success, than to take the lead in the introduction of a new order of things.[1]

Perilous or not though we must address this fundamental issue; laying out the china for yet another tea party to raise awareness about loss of the rainforest or organizing a drinks soirée to tackle famine in Africa is not going to solve the problem. The elephant stomping through our networking parties is our centuries-old, deeply corrupt and deeply corrupting, debt-based money system. That is what we have to change if we want a healthy planet.

If we believe that democracy is something worth having then we must make changes. We do not have democracy. Money is not the servant of the people. Money is a weapon used by those who control its creation and circulation. And that leads to the situation described by President Franklin Delano Roosevelt in his letter to Colonel Edward Mandell House:

> The real truth of the matter is, as you and I know, that a financial element in the large centers has owned the Government ever since the days of Andrew Jackson.[2]

We need people in all walks of life and in the rank and file of all political parties to recognize the need to create honest money. We can all, individually or as members of our chosen political parties, then fight our corners and debate all the issues we need to address: sensible and necessary business, healthcare, education and of course care for our environment.

We cannot have sensible debates on any of these issues while we allow a corrupt money system to make all the real decisions.

Conclusion

I have tried to make a case that we *Homo sapiens* (wise man) are not quite living up to our name. I have selected five areas where we could improve our performance. I will assume if you have got as far as this in the book then you are in general agreement with me.

I have tried to show that there has to be something fundamental that is at the root of our lack of performance in these areas. That 'something' is money. Money should be one of our most powerful and useful tools but because of the way we operate our monetary system it is our most powerful and destructive tool.

The evidence is all around us.

The Environment

We have an environment that is under threat but tackling the problem is impossible while we ignore the herd of elephants in the room. Jeremy Clarkson and Jonathon Porritt can't sit down and have a sensible debate about the best way forward because, whatever they agree to, the economy will ignore; it will demand growth at any cost no matter what any of us say or do.

Poverty

Poverty is all around us in our global village. What would we think if we were told that the Apache or the Cheyenne had allowed a few score to live with reasonable dignity? That they allowed half a dozen chiefs to enjoy a hundred wigwams apiece, fifty horses and a thousand dogs. And all this was while thousands of their fellow tribesmen were living in squalor while they slowly starved. Would we applaud them?

War

It is still the same ghastly racket that Major General Smedley Butler described nearly eighty years ago. Consider our shock at watching those passenger planes flying into those buildings in the USA. Consider what we should feel when a country creates a plane that has the specific purpose of destroying buildings. Remember that Joseph Stiglitz says that three trillion dollars and two million men have been used to fight just one war against Iraq. No matter what the problem is, this is no way to solve it.

Spending Cuts

To make the SS Britannica into a seaworthy craft once more we have to reduce the footprint of education, care for the elderly, public transport and health provision. This will all be done so that we can increase the footprint of the car industry, the fizzy drinks industry, the cosmetics industry. And all because we must make any sacrifice necessary to appease the economy stupid and get it up and running at full charge again.

Boom and Bust

Of course it will happen again. It is mathematically impossible for the cycle not to happen again. The system demands it. The system only works if people are increasing the amount of debt that exists. We need money and all the money circulating in the world was created when somebody went to a bank and took out a loan. As the Credit Manager of the Federal Reserve Bank of Atlanta put it: 'Someone has to borrow every dollar we have in circulation, cash or credit . . . We are absolutely without a permanent money system. When one gets a complete grasp of the picture, the tragic absurdity of our hopeless position is almost incredible, but there it is.' And he was in deadly earnest because he added this comment to his warning: 'It is the most important subject intelligent persons can investigate and reflect upon.'

The evidence is there to see. These are five issues that we have to deal with. We cannot deal with these issues though if we ignore the cause. Recycling and carbon credits won't save the planet. Comic Relief won't end poverty. Diplomacy and demonstrations won't end war. Strikes and campaigns won't give us a healthy society. And regulations and virtue won't prevent the next crash.

We must have, as Robert H. Hemphill of Atlanta said, a permanent money system. At the moment in this country I would suggest less than one in ten thousand have even given it a thought. And with good reason as J. K. Galbraith said: 'The study of money, above all other fields in economics, is one in which complexity is used to disguise truth or to evade truth, not to reveal it.'

We have to move forward from one in ten thousand seeing through the 'complexity' which is disguising the truth to a position where closer to one in twenty are able to see through the disguise.

We need an honest review of our monetary system. The creation and circulation of money is far too important to be left in the hands of private individuals who look upon money as a philosopher's stone capable of manufacturing wealth out of nothing. It is quite simply the most important issue that we have to deal with.

We must resist the idea that the creation of money and its circulation is too esoteric to be of any real importance. It is quite definitely of the utmost importance.

We must end what Congressman Wright Patman called 'this idiotic system'. We must finally change what Thomas Edison called 'this terrible situation when the Government, to increase the national wealth, must go into debt and submit to ruinous interest charges at the hands of men who control the fictitious values of gold'. We must make sure that the people really do 'have the benefit of their own gilt-edged credit'.

This will not be easy but remember what Jared Diamond said: 'We are the cause of our environmental problems, we are the ones in control of them, and we can choose or not choose to stop causing them and start solving them.'

Unfortunately 'we' are not in control. Our debt-based money system is in control. Fortunately we can change it.

The solution is not complicated. As Diamond says: 'We don't need new technologies . . . we just need the political will to apply solutions already available.'

I hope I have persuaded you to look again at why we have not so far found that political will. After all, as Francis Bacon said, knowledge is power. It is certainly the first step on the road to it.

Bibliography/Notes

Chapter 1
1. Jared Diamond, *Collapse: How Societies Choose to Fail or Succeed* (Penguin, 2005).
2. Thomas L. Friedman, *Hot, Flat, and Crowded* (Farrar, Straus and Giroux, 2008).

Chapter 5
1. Niall Ferguson, *The Ascent of Money: A Financial History of the World* (Allen Lane, 2008; Penguin, 2009).

Chapter 7
1. Graham F. Towers, Governor of the Central Bank of Canada (from 1934 to 1955), before the Canadian Government's Committee on Banking and Commerce, in 1939.
2. Robert B. Anderson, Secretary of the Treasury under Eisenhower, in an interview reported in *U.S. News and World Report*, 31 August 1959.
3. J. K. Galbraith, *Money: Whence it Came, Where it Went* (Penguin, 1975).
4. Thomas Edison, *The New York Times*, 6 December 1921.

Chapter 8
1. John Adams letter to Thomas Jefferson, 25 August 1787.

Chapter 9
1. Robert Peston, *Who Runs Britain? . . . and who's to blame for the economic mess*

we're in (Hodder and Stoughton, 2008).
2. J. K. Galbraith *The Affluent Society* (Hamish Hamilton 1958; Pelican Books, 1962).
3. Robert H. Hemphill, Credit Manager of the Federal Reserve Bank of Atlanta, in the Foreword to *100% Money* by Irving Fisher (Pickering Masters Series 1934).

Chapter 10
1. Samuel Huntington, *International Security* 17, no. 4, 1993.
2. Major General Smedley Darlington Butler, *War is a Racket* (Round Table Press, Inc., New York, 1935).
3. Noam Chomsky, *Hopes and Prospects* (Hamish Hamilton, 2010).
4. Emperor Hirohito of Japan, Imperial Surrender Broadcast, *August 15 1945.*
5. Reichsmarschall Hermann Goering in a conversation with U.S. Army Captain Gustave M. Gilbert – a German-speaking intelligence officer and psychologist, 18 April 1946.
6. Vince Cable, *The Storm* (Atlantic Books, 2009).

Chapter 11
1. Michael Chossudovsky, *The Globalization of Poverty and the New World Order* (Global Research, 1998).

Chapter 12
1. Nicole Gelinas *After the Fall: Saving Capitalism from Wall Street – and Washington* (Encounter Books, 2009).
2. Carmen M. Rheinhart and Kenneth S. Rogoff *This Time is Different: Eight Centuries of Financial Folly* (Princeton University Press, 2009).

Chapter 13 1. Vince Cable questioning Gordon Brown – Hansard, 13 November 2003.

Chapter 15 1. Thomas Jefferson letter to John Taylor, 26 November 1798.

2. Bernard Lietaer, economist, one of the designers of the Euro, *Yes! A Journal of Positive Futures*, 1997.

Chapter 16 1. Niall Ferguson, *The Ascent of Money* (Penguin, 2009).

2. Congressman Wright Patman, Chairman of the House Committee on Banking and Currency from a speech in Congress on 29 September 1941.

Chapter 17 1. Nicolo Machiavelli, *The Prince* (published 1532 (distributed in 1513, using a Latin title, *De Principatibus*)).

2. Franklin Delano Roosevelt in a letter to Colonel House, dated 21 November 1933.